to P.H.N.

The Roots of Mankind

The Roots of Mankind

John Napier

LONDON: GEORGE ALLEN & UNWIN LTD

ISBN 0 04 573007 5 cased
     0 04 573008 3 paper

Printed in Great Britain
by Compton Printing Ltd., London & Aylesbury

Frontispiece: Author with footprints. (Courtesy of the Ealing Corporation.)

Acknowledgments

It is difficult in a few lines to express the debt one owes to one's colleagues even if it could be assessed. How does one disentangle ideas put forward by others in seminars and private discussions from those that one likes to think of as "original." To credit my senior colleagues, contemporaries, and past students with all that is their due in the following chapters is quite beyond my capabilities. Like Topsy, many of the ideas and points of view have "just growed" and to trace them to their source would be well nigh impossible. Fortunately most people understand this dilemma well. My primary acknowledgment, therefore, is to all my colleagues who have advanced my understanding in the fields of primate biology and human evolution.

There are many categories of indebtedness, the inspirational, the practical, and the economic and it is perhaps invidious to distinguish them, so to the following I would like to record my great appreciation for help of one sort or another that they have given me:

Miss Mollie Badham and Miss Natalie Evans of Twycross Zoo Park; Dr. Sam Berry; Mrs. Audrey Besterman, who drew most of the original diagrams in this book; Dr. Michael Blake, the Boise Fund of Oxford

University; Professor Ruth E. M. Bowden, Sir Wilfrid Le Gros Clark, Sonia Cole, Professor Peter Davis, Dr. Michael Day; Miss Barbara Dickson, my administrative assistant at the Unit of Primate Biology in London; Miss Frances Ellis of the Photographic Department at the Royal Free Hospital School of Medicine; Dr. Robin Fox, Dr. Lewis Gazin, L. V. Grayson, Esq., Dr. Colin Groves, Dr. Charles O. Handley, Jr., Dr. W. C. Osman Hill, Dr. Clifford Jolly, Dr. L. S. B. Leakey, Dr. Rainer Lorenz, Dr. Desmond Morris, Prue Napier; Miss Dorothy Osborn, my secretary at the Office of the Primate Biology Program, Smithsonian Institution; Dr. David Pilbeam; Dr. Ted Reed, Director of the National Zoo, Washington, D.C.; Dr. Suzanne Ripley, Dr. Michael Rose, Dr. Adolph Schultz, Dr. Elwyn Simons, L. G. Smith, Esq., sometime Headkeeper of the Monkey House of the London Zoo; Mr. and Mrs. Eric Sorby, who are responsible for many of the photographs of living primates included in this book; Dr. Richard W. Thorington, Jr., Professor P. V. Tobias, Dr. Alan Walker, Lita Osmundsen, and the Wenner-Gren Foundation for Anthropological Research.

Naturally, the opinions expressed in the book do not necessarily reflect the opinions of the authorities mentioned above.

Contents

Introduction

The Friends of the National Zoo (F.O.N.Z.), Washington, D. C., must shoulder part of the blame for the appearance of yet another book on human origins. The substance of the following chapters was delivered to them in six weekly installments early in 1969. Their kindly reception and their obvious deep interest in the subject matter prompted the not un-willing author to expand these talks into the chapters of the present volume.

The venue for the F.O.N.Z. lecture series was the Elephant House at the National Zoo and thus the audience was supplemented by some three dozen assorted non-primate mammals including elephants, giraffes, rhinos, and pigmy hippos, which received the lectures with a variety of sound effects ranging from the simple grunt (of approval?) to the frankly derogatory belch—at often singularly appropriate moments. I can-not claim that this reception affected my decision to write this book one way or another but it certainly moved me to write the essay "A giraffe's eye view" which comprises the first chapter.

Books on science can generally be divided into two kinds: textbooks or technical monographs directed towards professional scientists, and "popular" books for the edification of the "informed layman" as the phrase goes, though what he is informed about nobody ever specifies. This duality is presumably a hangover from times past when science and the citizen were poles apart; when the mystique of science constituted an impenetrable barrier between the in-people and the out-people; when scientists were demigods who spoke another language and breathed the rarefied air of the scientific Valhallas of the 20th century—the M.I.T. in Cambridge, Massachusetts, the Cavendish Laboratory in Cambridge, England, Mount Palomar and the White Sands testing grounds of Project Manhattan. But today we have Cape Kennedy where second-by-second commentaries from blast-off to splash-down are shared by scientists and laymen alike. Through the mass media, the general public of today is in touch with scientific advances that 30 years ago would hardly have penetrated beyond the closed doors of the Royal Society or the National Academy of Sciences. The jargon of science technology has become the cliché of newspaper and television science-correspondents. This is all very right and very proper. Science costs us all a packet and the least we deserve is a share in the fun.

The way this book has turned out is that it is neither a scientific textbook nor a layman's guide to human evolution, but something of a cross between the two. I think it will suit the needs of the *non-anthropologist* whether he be a biochemist, a sociologist or a real-estate man. The world of science is so diverse that an expert in one branch is hardly more than a well-informed layman in another. There may be one language of science but there are many dialects, and the credibility gap between scientific disciplines is perhaps in even greater need of bridging than the rapidly shrinking chasm between science and the citizen. Thus, inasmuch as this book is directed at all, it is directed towards scientists and laymen alike who wish to know more about man and his biological background.

The chief problem, therefore, is to establish an acceptable level of communication. Technical writing is a form of shorthand in which major assumptions are made about the background knowledge of the reader and in which there is no pandering to the neophyte. Popular writing, on the other hand, makes equally sweeping assumptions but in quite the opposite direction. It assumes that the reader knows nothing and that therefore what he is *not* told won't hurt him; by its nature, popular writing must be somewhat dogmatic and doctrinaire with the inevitable result that half-truths are as common as nuts in May. The difficulty lies in devel-

2

oping some sort of compromise between the two styles that is both readable and informative for non-scientists, and serious and scholarly enough for scientists who are not anthropologists. I am aware, however, that such a hybrid may not please everyone. Fortunately there are some precedents for the dual purpose literary style and the two in particular that I have tried to keep in mind are the Friday evening discourses at the Royal Institution (R.I.) of Great Britain and the articles in *Scientific American*. At first sight these institutions form an unlikely twosome, but while some 50 years or so separate the dates of their respective foundations, both institutions are devoted to the principle that a knowledge of science is a human right. In the words of Count Rumford in 1802, the founding father of the R.I., the function of such an institution is to foster a "general diffusion of an active spirit of inquiry and useful improvement among all ranks of society." As far as possible I have tried to maintain a level that would be acceptable at both 21 Albemarle Street and 415 Madison Avenue. If anything, I have favored the more traditional style of the Royal Institution (including citations in the text and the harmless "academic joke" so beloved by British scientists but so rigorously eschewed by their American counterparts) rather than the streamlined, staccato, though eminently readable, convention of *Scientific American*.

W. S. Gilbert, the librettist of Gilbert and Sullivan, once wrote, "Man however well-behaved at best is only a monkey shaved." This, in short, is the message of this book—but without the overtones of ridicule of Gilbert's aphorism. It is an old-hat message, perhaps, and one that has been the subject of numerous books and essays since Darwin first promulgated the theory of evolution, but it bears constant re-telling as our knowledge of the *facts* of man's evolution continues to accumulate. In the last twenty years more fossil facts about the early history of man have been discovered than in the preceding half century. In the last 20 years, too, in quite another field of investigation, the discipline of comparative animal behavior, or ethology, has grown a new arm whose function it is to determine the relevance for human culture of the behavior of man's zoological kin, the apes and monkeys. The scientific success and popular appeal of this new arm of ethology has tended to distract attention from the anatomical, paleontological, and genetical contributions which, in the long run, human ethology must depend upon for its contextual framework. E. O. Wilson has recently taken up the cudgels on behalf of the science of genetics. In a recent address he stated, "Anthropology must encompass human genetics extending that science to generate a new discipline, anthropological genetics, the study of the heredity of the behavioral traits that affect culture."[1]

In this book I have taken the liberty of assuming responsibility for anatomy and paleontology. I have tried to provide an account of the inherited factors of the human constitution, the stages by which these characters evolved and the environmental forces that, acting through natural selection, brought them about. I am an anatomist by trade, not an ethologist, and whilst fully recognizing the critical role of behavior in evolution, I have not attempted to dabble unduly in matters outside my competence. There are, however, some aspects of behavior (aspects which I like to term first-order behavior, but which ethologists generally lump together somewhat prosaically as "motor patterns") which I regard as much my business as theirs. These are behaviors that are firmly linked to anatomical structure and arise directly out of it; they include locomotion, resting and sleeping postures, manipulation, feeding behavior, and vocal and visual communication. Traits that are less dependent upon structure and more obviously influenced by environment, such as social behavior and other aspects of primate, including human, culture, are discussed but at no great length. I am prepared to propound, in a limited sense, on "nature" but not on "nurture."

The study of primate biology and human origins for which this book provides a general introduction, is a synthetic subject inasmuch as its students, its principles, and its practice are derived from a variety of disciplines including zoology, anthropology, vertebrate paleontology, ethology, and human anatomy. Nevertheless, in spite of its multifactorial composition, the material it deals with is homogeneous. The Order Primates contains an array of animals, living and extinct, that display a clear evolutionary continuity from the lowest to the highest. The essential conformity of man with other primates in morphology, physiology, and serology is beyond question and constitutes the principal rationale for the widespread use of non-human primates as experimental animals in medical research. That a similar degree of conformity extends to many aspects of behavior is the contention of a number of zoologists, ethologists, and psychologists today, who are already applying the lessons of non-human pri-

---

[1] Quoted, with permission, from a paper by E. O. Wilson "Competitive and Aggressive Behavior" presented at the Man and Beast Symposium held at the Smithsonian Institution in May 1969 and to be published shortly.

mate behavior to such areas as child-development and education. Equally important is that the lessons of primate behavior, where they prove to be relevant, should be applied to the wider, national problems of racial integration, higher education, urbanization, civil violence, and the population explosion. In a still wider context, primate biology has something to contribute to international problems of cooperation, competition, and aggression. The importance of primate biology for the sociologist, the political scientist, and the politician cannot be over-emphasized. Nevertheless a word of caution would not be out of place. It is fashionable just now to condemn the neglect of the biological sciences by social scientists. That's as it may be. Undoubtedly the concept of culture as a man-made affair having little or nothing to do with genetic inheritance is deeply entrenched in anthropology. The vogue is to study man as a creature outside nature and inside a unique sort of cultural astrodome of his own creating, that owes little of its design to nature but much to nurture.

The opposing view is that the variables of modern human societies can all be traced back to the universals of primate behavior. Such reductionism is carrying the zoological perspective too far—and too soon. Human ethology, while paying lip service to Darwinian theory, sometimes fails to do justice to its principles. Some aspects of human culture, aggression for instance, need not be regarded as the stigmata of our primeval, lusty, animal past but as the genetic consequences of our lusty, animal present. Evolution is not an extinct phenomenon in man; it is still an ongoing force; therefore there is no reason to blindly assume that we are conveniently absolved of guilt by reason of the animal nature of our Pleistocene doppelgängers.

Leaving a high horse for a low one, with which I am much more familiar, I invite you to join a literary safari, a journey into the little-known world of primates, with man as the goal. To begin with we shall be looking at the Order in objective zoological terms in order to establish a base-camp from which we can pursue promising trails (on the track of man) with the knowledge that our lines of communication are secure. These trails may become rather indistinct and may even peter out here and there; but, pressing on in our search for man, making use of scattered clues, we can eventually reach our goal which, in the opinion of the author, happens to coincide with the conglomeration of lava and clay that comprises Bed I, Olduvai Gorge, Tanzania.

CHAPTER ONE

A giraffe's eye view

To a thoughtful giraffe the non-human primates might appear to be a rather uninteresting group of animals; generalized in structure, dull in color, unselective in diet, harmless in nature, tediously ubiquitous in distribution, and totally deficient in the well-known charisma of giraffes.

To an intelligent rodent, on the other hand, these negative qualities would appear as positive virtues. Primates, like rodents, are somewhat anonymous animals which have preferred the technique of steady unrelenting infiltration into the world, to the sensational entrances of certain animal prima donnas.

Can the accolade of harmless anonymity really be applied to the human primate, to man? Man, though undeniably a primate in a zoological sense, is but a single species among the 189 living species that comprise the Order; and it is with the characteristics of the Order as a whole, rather than the eccentricities of one of its species, with which primate biology is concerned. Furthermore man's recent psychosocial evolution would seem to have sequestered him so far from the realm of the non-human primates that even a thoughtful giraffe might well be excused for regarding man as a member of a completely different zoological Order.

Man, in every way, is a good primate. By the very possession of this knowledge, he exhibits the acme of a major evolutionary trend of the primate Order—an intense curiosity about things and events beyond his immediate ken. In man the desire to find out transcends the material inquisitiveness that leads monkeys to open up paper bags and compels young orangs to dismember any mechanical object that comes into their hands. Man's curiosity is conceptual and abstract and is concerned with such issues as religion, ethics, values, and the motivations of his own actions. His constant quest for knowledge about himself is his ultimate specialization which may, as the years go by—and this trait is more keenly developed by natural selection—lead to total autonomy in his own environment. The existence of a chance factor in evolution which has served for so long, though wastefully, as the principal mechanism of change may finally be superseded by genetic engineering, as it has been called, that will brook no aimless sorting. No happy-go-lucky system of random mating, which guarantees the universality of human genes, will be able to mar the perfection of computerized selection. Evolution will then, as it were, go out of business as a natural process in man, having evolved the seeds of its own dissolution. Curiosity or exploratory behavior will have become—as Desmond Morris (1967) percipiently calls it—"the greatest survival trick of our species."

The zoological strength of primates lies in the unspecialized nature of their structure and in the highly specialized plasticity of their behavior. This combination is hard to beat and has allowed primates to seize and exploit the wide variety of novel ecological opportunities that have come their way. The more specialized the primate, the more committed it is to a certain way of life and, thus, the more limited its potential for survival in a changing environment; and primates have been exposed, in the 65 million years of their history, to major environmental changes. Specialization is not always a liability. In the face of strong competition in a *stable* environment, it is advantageous; but at times of change it is the less specialized animals that have the edge over the more specialized. The primates, in the course of their evolution, have repeatedly been faced with changing conditions; a facility to seize new opportunities has been the keynote of their pre-eminence. Although there are a number of committed specialists among the primates (the gibbons, the aye-aye, the tarsier, and the potto, for instance), the hard core of the Order has maintained a position in the midstream of evolution, avoiding the drag and stagnation of life near the bank.

It is very questionable whether, in the context of vertebrates, the term "specialized" can be applied to a whole organism, or even to a whole organ. Man is unspecialized in possessing five digits on each hand, but is highly specialized in the possession of a broad terminal bone to his thumb. Taken as a morphological whole, the human hand departs less from the primitive pattern than the hand of a potto or a spider monkey; yet judged by its deployment in the field of human activity, it is highly specialized. The same problem can be looked at from a different point of view by substituting the organism for the organ. Baboons and macaques have evolved the most complex social systems among non-human primates. Physically, however, these animals are extraordinarily generalized. They possess relatively few morphological concessions to life in the savannas. Compared with other mammalian species that have adapted to a ground-living way of life, baboons are simply made-over arboreal monkeys and only distinguishable from them on the most trivial of structural grounds. It would appear that specialized behavior is not always associated with specialized structure.

The most successful primates in terms of population members and territorial spread *are those that have departed least from the ancestral pattern of structure but furthest from the ancestral pattern of behavior.* This on the face of it seems paradoxical, for structure and behavior are generally assumed to be interdependent. The key to this paradox is in the nature of the primate brain with its increased learning capacity and power of retention (Rensch, 1959). The brain of higher primates with its highly developed associative functions, its aptitude for receiving, analyzing, and synthesizing sensory impulses and converting them into appropriate, finely adjusted motor responses, forms the basis of a plasticity of behavior in primates by increasing their repertoire of possible responses to a given situation. The more generalized the structure of primates the greater is the range of their possible reactions and, thus, the more fitted they are to take advantage of new ecological opportunities.

The generalized nature of primates is nowhere better exemplified than in the suborder Anthropoidea, which comprises the monkeys, apes, and man. Within this category are three superfamilies, the Cercopithecoidea (Old World monkeys), the Ceboidea (New World monkeys), and the Hominoidea (apes and man). To the untrained eye, the typical representatives of the first two groups are indistinguishable; they are both "monkeys" that have short faces, forwardly facing eyes, flattened ears, five digits on each hand or foot, flat or flattish nails on the digits, and tails. Yet paleontological and zoogeographical evidence indicates that New and Old World monkeys have pursued independent evolutionary courses since the Eocene some 40-50 million years ago. Their common ancestry from a European or North American representative of the omomyid stock of prosimians has bequeathed them a common evolutionary potential. This and the similarity of their subsequent habitats has produced, after a vast period of geological time, two groups of monkeys so superficially similar that few visitors to a zoo seeing a New World monkey and an Old World monkey in adjacent cages, would acknowledge any major distinction.

Within the Cercopithecoidea, the generalized nature of the anthropoid suborder is even more apparent. Old World primates, like other groups, have undergone adaptive radiation during their evolution, but—unlike the prosimians, the suborder that contains the lemurs and the lorises—they have produced no really aberrant forms. Gibbons, baboons, gorillas, langur monkeys, and man show no fundamental, qualitative differences in the structure of the skeleton, alimentary system or central nervous system, nor are there any major differences in the microstructure of their tissues or the chemistry of their proteins. Such differences as are apparent are principally quantitative and concern the proportions of the limbs and trunk and the shape of the skull and jaws. These differences express themselves largely in the course of pre- and postnatal development as a result of relative growth rates.

In the phylogenetic succession from *Notharctus,* a lemuroid of the Eocene, through *Proconsul africanus* of the Miocene of East Africa, to modern man there has for instance been a remarkable conformity in the structure of the hand (Figure 1). Each has the same number of digits and the same number of bones in each digit. In man, however, the bones of the wrist show a numerical reduction from 8 to 7. The "missing" bone in the human carpus is the *os centrale* which occupies an intermediate position between the two carpal rows in *Notharctus* and *Proconsul africanus* and in all living primates. The uniqueness that this reduction would appear to give to the human hand is illusory because the *os centrale* is not really missing at all. It manifests itself as a separate cartilagenous element in the human hand well before birth and thereafter fuses with an adjacent wrist bone called the scaphoid. Fusion between the *os centrale* and the scaphoid also takes place in the hands of apes (and a few monkeys) at varying times after birth. It can be seen therefore that man is not unique in this respect except in the matter of developmental timing;

in respect of the *os centrale,* the human growth processes have been accelerated. With certain other human growth processes, however, the reverse phenomenon takes place and certain structures are developmentally retarded when compared with the same structures in other primates. For instance, compared with all non-human primates, the eruption of the second permanent molar in man is delayed until the rest of the milk dentition has been completely replaced. A further example of growth retardation during development leading to structural differences between adults, concerns the balance of the skull on the vertebral column. At birth, as Adolph Schultz, the doyen of comparative primate anatomy, has shown, the foramen magnum is situated at approximately the midpoint of the base of the skull in all primates. Postnatal growth in most monkeys and apes proceeds more extensively in front of the foramen magnum than behind it with the result that—with growth—the foramen magnum becomes more and more backwardly displaced rela-

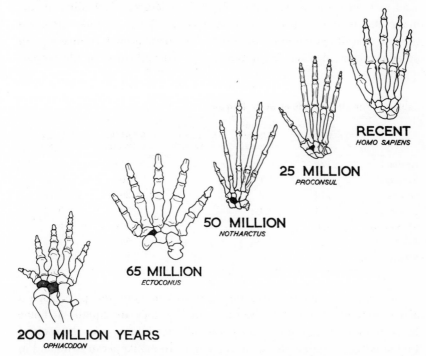

Figure 1. Evolution of the hand from a mammal-like reptile of the Permian to modern man. Shaded areas represent the *os centrale.* Note the conservatism in hand form from early Tertiary to the Recent. (From "Hands and Handles," by J. R. Napier. Courtesy of *The New Scientist,* 1961.)

tive to the base of the skull. In man growth rates in front of the foramen magnum and behind it are approximately equal and as a result the foramen magnum retains its central (fetal) position. In this characteristic, man—and certain other primates such as *Saimiri* (the squirrel monkey) — is said to display pedomorphism, a propensity for retaining fetal characters into adult life. Man, too, is pedomorphic in the development of the labia majora, the outer folds of the female vulva which appear in fetal and juvenile stages of chimpanzee life but disappear as adulthood proceeds (Hill, 1952). Man retains these folds throughout life. Contrary to the views of certain biologists such as Bolk, the protagonist of pedomorphism, who saw Man as merely a "fetal ape," the majority of man's characteristics are hypermorphic and not pedomorphic; that is to say, they are the result of the speeding up of growth processes and not of slowing them down.

Structural differences between man and apes, and apes and monkeys, tend to resolve themselves into differences of degree, depending on the mode and tempo of growth processes. Quite clearly in man there has been an overall reduction in the speed of postnatal growth compared with other primates; this has led, *inter alia,* to a lengthening of life periods (Table 1).

Table 1 **The Life Periods of Some Primates**

	Fetal Phase (days)	Infantile Phase (years)	Juvenile Phase (years)	Adult Phase (years)	Life Span (years)
Lemur	126	½	2½	11+	14–15
Macaque	168	1½	6	20	27–28
Gibbon	210	2	6½	20+	29–30
Orang-utan	233	3½	7	20+	30–31
Chimpanzee	238	3	7	30	40
Gorilla	265	3+	7+	25	36
Modern man	266	6	14	50+	70–75

Primates are highly adaptable creatures in both the physiological and anatomical senses of the word. Physically, they are equally at home in the trees, on the ground, on vertical cliffs, scrambling over the roofs of Indian villages and temples, or in the water. All monkeys can swim and in some species—the proboscis monkey of Borneo and the crab-eating macaques of the Philippines and other islands of southeast Asia for instance—water constitutes a major feature of their normal habitat. Apes, on the other hand, cannot swim naturally and for this reason are afraid of water, a fact that, with varying success, permits them to be exhibited on "island" enclosures at zoos. This is a strange difference in behavior which has never been fully studied. One assumes that it is connected with differences in the normal patterns of gait. Monkeys are quadrupedal and therefore only have to "walk" in the water in order to swim whereas apes which are arm-swingers during locomotion (brachiators) are required to learn a new pattern of movement to keep afloat. Man, too, has to learn to swim.

Even the most arboreal of monkeys have little apparent difficulty in adapting themselves to life on the ground; spider monkeys (*Ateles*) for instance, which seldom put foot to ground in their native Amazonian habitat, thrive in many zoos in treeless, concrete enclosures. This locomotor adaptability of primates has been of considerable importance to the Order as a whole during their evolution, permitting certain primate stocks to survive a major shift of habitat in the face of drastic environmental changes.

It is customary to label different stocks of primates in terms of their dietary preferences. There are generalized herbivorous monkeys which eat fruit, leaves, bark, flowers and buds indiscriminately; there are foliivorous species that, principally, eat leaves; there are insectivorous primates and there are carnivorous primates. But apart from a few specialized eaters like the colobines, the so-called leaf-eating monkeys, which possess a somewhat specialized stomach and large intestine, the vast majority of primates are omnivorous. Even the redoubtable gorilla, a notorious vegetarian, thrives in captivity on a regular supplement of meat. The presence of a generalized dentition and a generalized alimentary system in the majority of primates is evidence enough of their lack of specialization.

Primates today live in the tropics and subtropics. The vast majority are confined within the latitudes 25°N and 30°S, within the zones of rain forest, montane forest, and tropical savanna; but primates are individually adaptable and capable of adjusting their heat-regulating mechanisms to life in temperate and even cold climates. In the U.S.S.R., species of the genera *Macaca* and *Cercopithecus* have survived the outdoor rigors of Russian winters with no greater disaster than the loss of the tip of the tail. It is probable that primates, particularly the ground-living members of the Cercopithecinae (baboons and macaques), are now only restricted to the tropics because man has pushed back their frontiers to latitudes that as yet he has not been driven to exploit commercially. Were it not for the coming of man, macaques and baboons might well have been the dominant form of animal life in all the temperate regions of the Old World. In intelligence, adaptability, and behavioral plasticity, the macaque and baboons, which are closely related, are the leading non-human primates. If man is President of the animal world, then the macaque is certainly the Vice-President. There is no *a priori* reason why certain primates should not be maintained and bred successfully in temperate regions; primates are no more specifically adapted to the tropics than man is, and there is no reason to doubt that their homeostatic adaptability is not equally effective.

These three examples, locomotor adaptability, digestive adaptability, and climatic adaptability reflect the unique potential possessed by primates, a potential for opportunistic evolution that they have never failed to seize. The ability to enter and exploit new ecological niches as they become available has been the recurrent theme of primate evolution. This potential has been fully realized in the emergence of *Homo sapiens*. the quintessential primate, a product of 65 million years of physical non-commitment.

Giraffes see the world of primates from an average elevation of 18 feet. Man is inclined to see it from an altitude of 6 feet, more or less. Man sees primates in his egocentric way as "little men," which naturally loom large and important among mammals. To the giraffe, primates are a familiar part of the daily scene, harmless but rather restless creatures, typical run-of-the-mill mammals in fact, with four legs and a tail, big eyes, small ears, ordinary noses and pathetically short necks. Clearly there is nothing very extraordinary about primates compared with giraffes which, as everybody knows, are unique and very charismatic animals and are *so* extraordinary that even man, the primate of primates, was once heard to remark, on seeing a giraffe for the very first time, "There ain't no such animal."

CHAPTER TWO

Evolutionary theory and the meaning of species

The theory of evolution, the idea of change with time, did not start with Charles Darwin. Jean Baptiste Pierre Antoine de Monet, Chevalier de Lamarck (1744-1829), Erasmus Darwin (1731-1802), and Charles Lyell (1797-1875) had all anticipated the theory in some form or other; but it was Darwin who was able to synthesize the accumulated data of half a century, add the essential ingredient and present a coherent, wholly convincing theory that promulgated the principle mechanism of this change —Natural Selection.

Darwin successfully challenged the accepted ideas derived from Baron Cuvier (1769-1832), the arch catastrophist of the period, who believed from the evidence of the discontinuity in sedimentary rock formations that new waves of life, following on the heels of catastrophes such as floods and earthquakes, introduced a new faunal package of "improved" species, biologically unconnected with any previous, pre-catastrophic assemblage. This, the so-called theory of progressionism, was a "fundamentalist" doctrine of disjunctive evolution, neatly tailored to explain the undeniable evidence of the rocks, which revealed progressive change in fossil species.

The catastrophist theory was strongly upheld by Paley in his *Natural Theology* (1802), a book whose influence lasted well beyond the publication of *On the Origin of Species,* and provided the ammunition dump from which were derived the thunderous cannonades of Richard Owen and Bishop Wilberforce against the redoubts of Darwin's theory.[2] The "Bridgewater Treatises" reinforced the teleological philosophy of Paley. These were a series of monographs published under a trust fund left to the Royal Society by the Earl of Bridgewater. Among the best known of these was *The Hand—its Mechanism and Vital Endowments as Evincing Design* by Sir Charles Bell, published in 1820. Bell, like Cuvier and Paley, was a teleologist—a believer in the doctrine of final causes (i.e., "The giraffe has a long neck *in order* to browse off high branches"); also, like Cuvier, he was a great comparative anatomist who appreciated the importance of structural adaptation for animal survival. In his belief that the mechanism of adaptation was special creation, Bell was merely reflecting the most widely held scientific convictions of his time.

Lamarck was the archenemy of the catastrophist theory. He advocated the uninterrupted continuity of life, believing that species graded into one another by transformism (a sort of evolution) and that the whole process was a straight line sequence of increasing perfection and complexity leading from the lowest organism to the highest.

The greatest stumbling block of all those who sought to resist the doctrine of teleology was to explain adaptation. Progressionism, after all, provided an explanation: adaptation was the special creation of novel forms to fit the new, postcatastrophic, environment, and it behooved Lamarck to meet this challenge and to produce a new theory to explain the progressive transformation of one species to another. In essence the Lamarckian philosophy states that, as the result of the "needs" experienced by animals in their natural surroundings, new organs developed which, from then on, become part of the inheritance of the species. Lamarck's theories are inevitably wedded in the minds of students to the thesis of "the inheritance of acquired characteristics" but, as George Gaylord

---

[2] But in other ecclesiastical circles the astonishing news was received with commendable philosophy. The story goes that the wife of a Canon of Worcester on being told of Darwin's theory that man was descended from monkeys said: "Let us hope that it is not true, but if it is let us pray that it will not become *generally known.*"

Simpson points out, Lamarck did not believe that *all* acquired characters are heritable—he held that only *some* of them could be passed on in this way. In particular he was concerned with the inheritance of behavior. Lamarck's major contribution to biology was that, following Buffon, he was the protagonist of a scientific theory to explain the evolutionary continuity of life.

Charles Lyell, the author of *Principles of Geology* (1830), was, according to Eiseley writing in 1959, the single most important influence in Darwin's life. In 1831 when Darwin set out in H.M.S. *Beagle,* he carried with him the first volume of Lyell's book; the second volume reached him *en route* according to De Beer (1963). To Lyell, Cuvier's theory of catastrophism and its derivative, progressionism, were unacceptable. Lyell, like Lamarck, believed in the continuity of life, but, at the same time, rejected Lamarck's view that animal life constituted a continuous spectrum of intergrading species. Lyell knew from his paleontological studies that certain species had become extinct and had been *replaced* by other species, but the process by which this could have come about, completely evaded him. He was aware of the "struggle for existence" and indeed wrote about it. He also knew about natural variation and niche adaptation but he could see no means by which new forms could emerge and survive in an adaptive zone already occupied.

It is in no way to belittle Charles Darwin's great contribution to biology to say that the seeds of evolutionary theory had already been sown by the time Darwin, fresh from his extraordinary adventures on H.M.S. *Beagle,* brought his original mind to bear upon the problem. Eiseley (1956) rightly stresses the genius of Darwin when he states "Almost every great scientific generalization is a supreme act of creative synthesis that represents the scientific mind at its highest point of achievement." Natural selection, the real missing link in the evolutionary theories of the 19th century, was forged by Darwin and with this in place the chain was complete. The theory of evolutionary change through natural selection, as envisaged by Darwin, was based on his observations on natural variation, variability in animal populations, and the need for a natural force to control the size of animal populations—the latter was a critical aspect of his theory which he learned from the writings of Malthus. Darwin made the assumption (since proved conclusively) that some inherited variations are of survival value to the animal or plant possessing them. Therefore, individuals with these advantageous characters are likely to be the ones that breed *more* individuals of the next generation than those without such advantages; the next generation therefore will have a higher proportion of individuals possessing these advantageous traits than the preceding generation. They will show a greater "fitness." As this process continues, the population will eventually contain only individuals with these characters, and if they happen to be major characters that lead the group that possesses them into new adaptive zones then, perhaps, a new species will evolve. To Darwin (ignorant of the work of Gregor Mendel, an Austrian monk, on inheritance of genetic traits and of the whole science of genetics which stemmed from it) the chief weakness of this assumption was that it depended on the ready availability of variation within the population, the mechanism of which was not apparent to him. Intraspecific variability is now accepted as the principal source of evolutionary novelties, the basic mechanism of morphological and (some) behavioral changes.

DEFINITIONS OF SPECIES

Since the days of Linnaeus (1707-1778), species have been regarded as the basic units of the classification of animals but, until recently, they have been generally regarded as arbitrary, morphologically defined, static categories. The first step that led to the new "species concept" was the recognition of polytypic species or species that contain a number of geographic races or subspecies. The recognition of geographic races, having populations with "intermediate" characters, provided the evidence that evolution was an ongoing affair, continually modifying the type-characters of the species. The idea of change with time thus began the erosion of the historic concept of immutability.

Since the times of Linnaeus, morphology has been the principal criterion by which species have been defined. The difficulties inherent in such definitions are that "good" species (species that can be unequivocally shown to *be* species) may not differ significantly in anatomical detail from each other; nor, on the other hand, do morphological differences in all instances indicate that two populations really constitute two distinct species. *Pithecia pithecia,* the white-faced saki monkey from South America, for instance, shows a marked difference between the sexes in external characteristics. For many years male and female sakis were placed in separate species. Apart from these special cases, however, the degree of difference between species is still most usefully (and most commonly) demonstrated in anatomical terms; species that are so distinguished are called morphospecies. The limitation of the anatomical approach to species definition is that it cannot be used as means of separating species absolutely. This process must always contain an element of arbitrariness, a process of drawing-the-line-somewhere. The difficulties in using morphology as the sole indicator of difference is that most characters are quantitative in nature, that is to say they are expressions of continuous variation within a spectrum of size and shape (see p. 23). There are few real qualitative characters of difference between members of the same family, let alone the same species. Many characters that in the past have been proposed as qualitative have been shown to be strictly quantitative in origin. Quantitative characters are no respecters of taxonomic categories and freely cross specific, generic, and even family borderlines. The laws of nature, the physical laws concerned with the response of animal tissue to size, to gravity, to temperature, to injury, and so on, operate on all organisms, tiresomely irrespective of taxonomic considerations. Allometry, the principle according to which the proportions of animals change with size, is a potent source of confusion in the taxonomy of closely related animals; structures such as the brain, the jaws, and the limbs are particularly subject to allometric growth effects. Species can only be "roughed out" on the basis of morphology; other criteria are needed to prove or disprove conspecificity.

Species have been defined on the basis of fertility. The production of fertile hybrids has long been regarded as evidence that two breeding populations belong to one and the same species. There is little justification for this belief, as many instances of interspecific and even intergeneric crosses are known. For instance, species of the genus *Macaca* interbreed readily in captivity. Crosses are also known between primates of different genera: macaques and baboons, baboons and mandrills, mangabeys and macaques. Although interbreeding in the wholly artificial environment of the zoo and laboratory is well recognized, these sort of crosses seldom occur in nature for a number of very obvious reasons. Firstly, the areas of distribution of the animals do not always overlap and therefore they do not meet; and, secondly, even if they did meet, courtship and mating behavior is so variable between genera (even between species) that it is highly improbable that the love affair would even get to first base, let alone reach a mating point. Finally, for interfertility there has to be a compatibility of sexual anatomy.

Although many species of primates are reproductively compatible they are, nevertheless, reproductively isolated by a combination of geography, ecology, social behavior, and morphology.

In 1940, Ernst Mayr of Harvard, a highly regarded taxonomist, provided a definition for species which omits any reference to morphology or experimental "crossability": *Species are groups of actually or potentially interbreeding natural populations, which are reproductively isolated from other such groups.*" Thus, species are objective, non-arbitrary categories, that are freely observable in nature.

Simpson (1961) has criticized this definition on the grounds that it constitutes a special case of what should be a broader evolutionary concept. Simpson's preferred definition runs as follows: *"An evolutionary species is a lineage (an ancestral-descendant sequence of populations) evolving separately from others and with its own unitary evolutionary role and tendencies."*

These two definitions, the neontological (Mayr) and the paleontological (Simpson) are clearly quite compatible.[3] Both classifications are equally phylogenetic in approach but are not equally applicable in practice. Whereas the taxonomist may be able to test the actuality or even the potentiality of interbreeding between two populations on the basis of field studies, he must depend on inference from fossil evidence—very often of the most incomplete nature—to determine the separateness of his evolutionary lineages. From such evidence as exists on the geology, botany, and climate of the past, he must infer a temporal, geographical, or ecological separation. It is apparent why Simpson's definition must rest, essentially, on morphological evidence—there is practically no other kind to go on in paleontology. Mayr, on the other hand, has cut to the minimum the inferential components of his diagnosis of species. In all equity, it must be pointed out, however, that actual or potential interbreeding is something that cannot always be determined, particularly in allopatric (geographically separate) populations without the existence of geographically intermediate, hybrid populations. Cain (1960) has attempted to solve this problem by suggesting that breeding populations that are within "cruising range" of each other should be regarded as potentially interbreeding.

In summary, it would seem that neither definition should be regarded as being the more "correct." They represent different ways of looking at the same thing and, as such, merely reflect the principal fields of interest of these two leading authorities in the field of vertebrate taxonomy.

---

[3] *Neontology:* The study of contemporary forms. *Paleontology:* The study of fossil forms.

A MODEL OF SPECIES EVOLUTION

The evolution of a species is illustrated in Figure 2 in a simplified and stylized form. Bearing in mind Ernst Mayr's definition, reproductive isolation (in effect, genetic separation) is the keystone to the formation of a new species. The diagram shows the fragmentation of an original population into two demes, units of population within a species, that remain within "cruising range" of each other, occupying different parts of the same geographical and ecological niche. The gene flow between these populations is reduced because, although there is no actual reproductive barrier, the opportunities for reproduction between the two demes are diminished by virtue of lack of propinquity. Consequent upon the occupation of two areas in which, inevitably, ecological conditions are dissimilar, minute structural and behavioral differences appear in the two demes.

At this point in the model, an agent of reproductive isolation is introduced. Agents of reproductive isolation are usually geographic. Whether or not speciation can occur *without* geographic isolation is a matter of present dispute. The agent in this model is a geographical feature, a broad river perhaps, a mountain range, a tongue of advancing glaciation or a desert. Now follows a period during which the two populations are wholly isolated in a geographical sense (allopatric) without an intermediate (hybrid) zone of intergradation. Genetic differentiation proceeds to a point when the two populations are sufficiently different, anatomically, to be regarded as separate geographic races or subspecies. Eventually genetic differentiation may reach such a pitch that the two geographic races are reproductively incompatible to such an extent that they fall within Mayr's definition of species.

The next stage in the model is the removal of the isolating mechanism so that our two species become once again geographically contiguous or even sympatric.[4] But even when this is done the behavioral reproductive barrier remains. Although two such species may still be interfertile, "crossable" in artificial laboratory experiments, the reproductive "barrier," is sufficient curb to reduce hybridization to insignificant proportions and to prevent the formation of a hybrid population. One tends to assume that the reproductive barrier is behavioral and that the "right" signals are not being exchanged, but the failure could also be morpholog-

---

[4] *Sympatric:* A term applied to populations within a single species that occupy identical or overlapping geographic zones.

ical. A Great Dane has a very real anatomical problem when attempting to mate with a Pekingese. The anatomy of sexual intercourse is species-specific. The length and the angulation of the erect penis is complementary to the depth and orientation of the vagina.

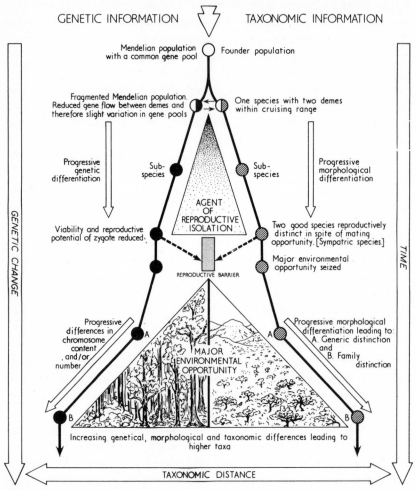

Figure 2. A model of the origin of species.

Figure 2 also provides a model for evolution of higher categories. There is no essential difference between the evolution of species and that of higher taxa such as families or genera. All evolutionary progression is a matter of natural selection operating on a pool of intraspecific variation. Macro-evolution is the appearance of major evolutionary *trends* leading to the production of higher categories and is largely opportunistic. It demands, firstly, an opening up of a new environmental opportunity, and, secondly, the presence of a group of animals in the right place at the right time. Furthermore such animals must be sufficiently plastic in behavior and generalized in structure to take advantage of a major, often novel, ecological niche. The invasion of a dry land habitat by the ancestral amphibians, the colonization of an aquatic environment by cetacean mammals and the adoption of flight by Jurassic reptiles are classical examples of opportunism which lead to the evolution of major taxonomic groups. Among primates two such major ecological shifts can easily be recognized, though no doubt there are many more less dramatic and less significant examples. The first was the adoption of an arboreal habitat by early Eocene prosimian primates and the second was the return to ground-living life during the Miocene by the hominid ancestors of man. These adaptive plateaus were critical events in the calendar of primate phylogeny and led to new evolutionary trends that have culminated in the diversity of the Order that embraces treeshrews, lemurs, lorises, tarsiers, monkeys, apes, and man.

SOME PRINCIPLES OF EVOLUTION

Variation is the difference in genetic structure within a population and is the source of evolutionary change. Without differences between individuals in a population, natural selection would have nothing to select from. As natural selection operates through differential breeding, there would thus be no available mechanism by which population growth could be controlled.

Variation appears in the individual animal in the form of traits which may be expressed in anatomical, physiological, or behavioral terms. These terms may be assessable either quantitatively or qualitatively. Stature in man provides a simple example of a quantitative variable. Differences in stature within a population constitute a continuous variable that can only be expressed, meaningfully, by statistical methods such as the normal distribution curve. It is, of course, perfectly possible to employ coarse units of classification such as "tall," "medium," and "short" by a process of arbitrary segmentation of the data as is customary for instance, when dealing with the colors of the spectrum. Other continuous variables relate to weight, limb-length, body types, tooth size, and skin pigmentation.

Qualitative variations are discrete and do not intergrade; in many instances they are either present or absent as, for example, the ability to taste certain chemical substances. Individuals are either "tasters" or "non-tasters" for a chemical substance known as PTC. The majority of humans find PTC obnoxious, but to others (20%) PTC is as bland as unsalted porridge. Other qualitative variables, much studied by anthropologists, are the antigenic factors in the blood, the ABO blood groups, Rh factors, and the M and N groups, etc. There are also a host of biochemical dysfunctions such as sickle-cell anemia, some forms of diabetes insipidus, and certain sex-linked diseases such as hemophilia that occur in some people, but not in others. Qualitative variations of this sort are of particular value to geneticists as tracers for testing genetic hypotheses, to demographers for plotting population movements, and even to historians constructing family trees. The so-called Hapsburg lip, the pouting lower lip of the Hapsburg royal line, can be recognized in contemporary portraits as far back as the 13th century. Porphyria, thought to be the true reason for George III's "insanity," is another heritable metabolic disorder. In this instance the disease has been traced back to his grandmother many times removed, Mary, Queen of Scots (Macalpine and Hunter, 1969) ; this interpretation, however, has been questioned.

In recent years variations have been described in blood proteins of non-human primates and the experimental evidence has been used to propound new theories of primate evolution. Although such methods are valuable adjuncts for the determination of phyletic relationships, it is very doubtful if they have much validity as isolated characters.

Anything that favors survival of an individual and promotes reproductive success ("fitness") is *adaptive*. Adaptations may appear in anatomical, physiological, or behavioral guise. *Natural selection* is the process by which traits, advantageous in a population, are perpetuated by the differential breeding capacity of its component individuals. Individuals which possess certain characters that make them more viable or fertile than other individuals in the population are likely to breed the greater number of offspring. The greater the number of offspring, the greater the chance that these new characters have of becoming perpetuated in a particular evolutionary line. Ultimately all genetically determined change that is not by virtue of its nature instantaneously lethal, must sooner or later face the test of fitness—reproductive success—if it is to have any influence on the course of evolution. The test bench for all genetic innovation is the environment, and *in this sense* one can say that the direction of evolution is determined by the environment. This is of course quite a different thing from saying that the environment *controls* evolutionary change. Environment merely has the last word in a process that starts as a mutation in the arrangement of the chemical substances of the DNA molecule and ends with the reproductive success of the mutant. However novel the adaptation, it is of no significance to future evolution if it cannot be transmitted to the next generation. Natural selection operates *via* the environment to foster characters which will provide the species with reproductive advantages.

In one sense, the whole way of life of an animal is a preparation for reproductive success. It has been said that locomotion is simply nature's way of approximating the male and the female of the species for reproductive purposes. This aphorism, though banal, has something to recommend it as we shall see in Chapter 5. In the same spirit of banality, it is certainly true to say that the introduction of the bicycle so increased the cruising range of the young men of the remoter villages of England that villages—with their tendency towards inbreeding—were dramatically changed in a matter of a few decades. Just as the evolution of the incest taboo in our remote Pleistocene ancestors led to exogamous mating and, thus to the foundations of human society, so the bicycle helped to raise the fitness (in an evolutionary sense) of the villagers of Old England— and no doubt of New England too. Hybridization between endogamous populations is to be rated as a good thing. In plants, at any rate, "hybrid vigor," or improvement of the hybrid over parent populations, results from such crosses, although whether or not this phenomenon operates for man is not known with any certainty.

Natural selection is not always a force for change; it is also a force for conservation. Less well-adapted organisms are constantly being eliminated by selection to maintain the *status quo* in a particular environment; this function might be called *natural rejection*. A good example of the stabilizing function of natural selection is the maintenance of the optimum birth weight of human infants. Babies that are too small or too big at birth are more prone to be "eliminated" for one reason or another in the immediate postnatal period or early in infancy than babies of optimal weight. The optimal weight for survival is reflected in the "average" weight of viable infants. One of the limiting factors in size of human infants at birth (and therefore one of the factors of the environment through which natural selection operates) are the proportions of the female pelvis. The birth of a human infant is, in the words of the Duke of

Wellington describing the battle of Waterloo, "the nearest run thing you ever saw in your life." The fronto-occipital diameter of the infant's head which is the part that usually presents at the vaginal opening, is as firmly held as a cork in a bottle. The fit is precise; if the head were any bigger or the pelvis any smaller, normal birth would be impossible. There would seem to be at least two adaptations that would solve this problem and make the birth process easier than it is at present; firstly, the baby's head could be smaller or, secondly, the pelvis could be larger. Let us consider the latter possibility here and leave the former to be discussed in a later chapter.

The proportions of the human pelvis are governed by a set of limiting factors which relate to human walking. The female pelvis is as wide as it can be consistent with the human type of gait; any further widening would lead to an imperfect, physiologically uneconomical, type of walk. This example not only emphasizes the importance of the stabilizing effect of natural selection, but it also illustrates that most selective effects are compromises between conflicting needs of the organism. Adaptations leading to the evolution of the human-type pelvis are compromises between the needs of childbearing and the demands of the upright posture of bipedal walking. It is perhaps not surprising that the human female is prone to develop pelvic defects such as femoral hernia, uterine retroversion and uterine prolapse, direct results of a compromise which, while invaluable during childbearing, constitutes a liability at other times. There are quite a number of other defects which plague the human race that can be attributed to man's specialized posture and gait (see Chapter 8).

Another example of adaptive compromise is now well recognized in the case of the sickle-cell gene. In certain parts of the tropics where malaria is common, the so-called sickle-cell gene is also common in the population. This is not a coincidence but an adaptation that comes about in the following fashion: individuals that have inherited the sickle-cell gene from both parents are said to be homozygous for this trait. The gene produces an abnormal hemoglobin (S-hemoglobin) which responds to lack of oxygen in a peculiar way that distorts the shape of the red blood corpuscles, which are then more rapidly destroyed by the body than they would be were their shape normal and not "sickled." Loss of red blood cells leads to the individual becoming severely anemic and death usually occurs early in life before the reproductive age is reached. Strictly speaking this should lead to elimination of this gene by natural selection but this has not happened. Clearly therefore the sickle-cell gene carries some kind of advantage. Individuals that have inherited a sickle-cell gene from one parent only are said to be heterozygous for the gene. They possess both normal (A) hemoglobin and abnormal (S) hemoglobin. In 1954 it was discovered that heterozygous individuals in regions in Central Africa where malaria is rife have a 25 percent better chance of survival than individuals with no sickle-cell gene at all. In some way this gene protects against malaria, and as it increases the fitness of its carriers the gene is perpetuated by natural selection.

Many expressions of adaptive compromise are met with in primate biology; particularly obvious are those where single structures have dual functions. The hand of higher primates is adapted for prehension and manipulation but this does not prevent it from being used as a foot during locomotion. The color of the coat of monkeys can probably be regarded as a compromise between the conflicting needs for concealment and recognition. The specialized, procumbent incisors of the lower jaw of lemurs constituting the so-called "dental comb," have a digestive as well as a fur-grooming function.

Since behavior has become a major subject of concern to biologists, there has been a tendency to take units of behavior and to fractionate them, apportioning so much to the genes and so much to learning. For instance it has been shown that some nestling song birds, reared in isolation, know *how* to sing (the subsong), but not *what* to sing (the species-characteristic song). This they only learn when exposed to other members of their own species. Isolation studies have shown that macaques and chimpanzees, deprived at birth of contact with others of their own species, can socialize when reunited with their peers, but they socialize imperfectly and are sexually naive. These studies have been quoted as examples of the nature-nurture dichotomy. A bird can sing (genetic) but lacks the syntax of song (learned); a male chimpanzee can approach a female sexually (genetic) but is unaware of what to do next (learned). However, singing and song-syntax, the urge to mate, and the performance of mating are not simply respective discrete sections of singing and mating behavior that, when tacked together in sequence, produce whole units of activity; rather they are reflections of two distinct *levels* of activity that go to make up a behavioral trait. Both levels are comprised of inherited patterns of behavior and patterns which are learned by experience. Even learning by experience is conditioned by genetic coding through the medium of the central nervous and endocrine systems. For instance, imprinting, the phenomenon first described by Konrad Lorenz in greylag geese, is inherent only in the nature of the gosling's "innate disposition to learn" as Tinbergen (1951) has called it. Imprinting is a learning process whereby the young bird fixates on the first object that it sees immediately after hatching. The gosling continues to behave towards this object, even into adult life, as if it were in effect its mother. If the object happens to *be* its mother, that is as it should be; but if it happens to be Konrad Lorenz (with the respect due to a great pioneer of ethology), that is as it should be too. The point I wish to make is that the innate component of this response is the capacity to latch on to the first moving object that comes within its ken; to learn in other words. The gosling carries no genes for correct mother-recognition. If it did, it wouldn't make such a silly mistake.

To take another example that is closely related of the difficulty of disentangling what is innate and what is learned, the male baboon has both an aggressive temperament (hormonal and innate) and long cutting canines (structural and innate). It inherits both the hormones that trigger off an aggressive response and its canines. But it has to learn throughout the long and important period of adolescence how these legacies can be deployed to the best adaptive advantage for troop survival.

Behavior whether it is genetically coded or socially learned is subject to natural selection; it is adaptive and therefore part of the fitness of the animal. The time-honored question whether behavior is a product of "nature" (innate or genetical) or "nurture" (learned or environmental) is really irrelevant as Dr. Glen McBride of the University of Queensland has recently pointed out at the Symposium on Man and Beast at the Smithsonian Institution. Irrelevant, because no simple nature-nurture partitioning exists; we are asking the wrong question. Learning, which is usually equated with the "nurtural" side of the nature-nurture dichotomy, is an adaptive process of the central nervous system and its effectiveness depends on the ability to learn—a trait which is itself genetically coded. There is no doubt, however, that learning can be reinforced by training (the learning-to-learn concept of animal behaviorists), but there is no means of distinguishing the inherited elements from the learned elements in any given unit of behavior; nor is there any particular reason for doing so.

Like all adaptive characters the capacity to learn and the deployment of what is learned is subject to considerable variability within the species. Cole (1963) describes a ranking order in a group of pig-tailed macaques while they were under laboratory study for visual discrimination tests. He noted that the monkey at the bottom of the class in "learning to learn" was at the top of the class in tests of manual dexterity. Jay (1965) reported on the variability in skill of female langur "aunts" when handling newborn infants belonging to other females. Many other examples of intraspecific behavioral variability are to be found in the literature of primate field-studies.

Another aspect of evolutionary change, which is intimately related to the appearance of new major ecological opportunities is *adaptive radiation*. In its most dramatic forms it is explosive, the original population rapidly diversifing into a number of different adaptive types. The rapid diversification of flowering plants (Dicotyledons) during the Tertiary led to the appearance of dense, stratified, closed canopy forests and to the opening of a whole new series of ecological niches above ground level. The colonization of nonforested areas by grasses and herbs in the early Miocene made possible the important savanna radiations of ungulates, carnivores, and primates. Faunal invasions into unoccupied areas following the opening of land bridges resulted in the rapid exploitation of new ecological niches. Although all adaptations within a group are radiating, in the sense that all animals concerned acquire a variety of differing characteristics, it is the speed and degree of the diversification that singles out adaptive radiations for special consideration.

Pre-adaptation indicates the possession of characters, physiological or behavioral, that given an appropriate ecological opportunity can be brought into play. One of the most important examples of this in the evolutionary history of the primates is that the arboreal adaptations of tree-living forms are pre-adaptive for life on the ground. Holding the body upright, an essential prerequisite for upright bipedal walking in man, was, in the first instance, an adaptation to life in trees. While pre-adaptation is a valuable concept it must be clearly appreciated that it is not an evolutionary mechanism *per se*; it is simply an adaptation that—in retrospect—appears to have played an anticipatory role. To imply an inherent purposiveness in pre-adaptation would be quite incorrect.

There is, however, another interpretation that can, perhaps, be made. Konrad Lorenz has emphasized that behavioral movements in birds tend to precede phylogenetically, the specialized structures that make such movements conspicuous. For instance, head-bowing in courtship behavior is accompanied in some species by the presence of specialized head crests that serve to draw attention to this movement. It seems possible that the behavior that Lorenz describes, which is so closely allied to the act of reproduction, would be favored by natural selection to a much greater extent than would some other, less sexually immediate, behavior and would be thus assimilated into the genetic code, a process which has been well demonstrated in laboratory experiments with fruit-flies (*Drosophila*).

The pre-existence of a particular behavior would put a high selection pressure on any physical trait, however small, that enhanced its effect. In view of the prominent role that learned behavior plays in the life of higher primates, behavioral pre-adaptation has undoubtedly played a large part in human evolution and probably underlies man's acquirement of bipedalism and manual dexterity.

A clear distinction must be made between the two processes of *adaptation* and *adaptability*. Adaptability is the function of an individual; it is a physiological or behavioral phenomenon by which the body adjusts *during its lifetime* to a wide variety of internal and external factors tending to discompose it. Adaptation on the other hand is a process initiated at genetic level, which increases the fitness of a particular population for a particular environment.

Individual adaptability is directed towards the neutralization of threats tending to disturb the steady state of bodily functions. Homeostatic processes usually imply physiological adjustments such as those involved in regulating body temperature, in maintaining adequate concentrations of chemicals in the blood, in drug habituation, and antibody formation. The concept of homeostasis can be extended to include behavioral mechanisms in addition to the obvious physical demands of the environment to which all warm-blooded animals must adjust themselves. For instance, one form of physiological adaptability that goes on in the central nervous system is called learning. Psychological adaptability is also a factor. Higher primates (and man in particular) have added a new dimension of sociality to their living. The social and cultural elements of a society are just as much part of the environment as geographical features. It is quite as important for survival that man should be able to adjust psychologically to changes in the social and political beliefs of his time as it is for athletes to compensate physiologically for changes in barometric pressure at high altitudes. Needless to say, adaptability is just as much a matter of genetic inheritance as is adaptation. This may sound like drawing a distinction without a difference. The difference is largely a matter of usage.

With the rejection of special creation to explain evolution, the principle of *orthogenesis* (or oriented evolution without the benefit of natural selection) became a popular substitute. The advantage of a scientific name to designate a metaphysical concept of evolution, with man as its cosmic goal, and the comfortable feeling of a compromise that it induced, was no doubt highly satisfactory to many 19th century scientists.

Orthogenesis must be distinguished from *orthoselection*, which is also a concept of oriented evolution but one based on Darwinian principles; it is characterized by a steady, progressive improvement occurring in an evolutionary lineage and presupposes some constant factor, such as the environment, operating to limit diversification.

Orthoselection has characterized the evolution of the Anthropoidea which throughout their phylogenetic history have undergone a grade-by-grade improvement with a minimum of diversification. This process has been given the name of *anagenesis* (Rensch, 1959).

A special case of orthoselection that has some important implications for primate phylogeny is referred to as Cope's Rule. Cope's Rule states that there is an evolutionary increase in body size within the lines of descent of many animals that has been established for many different groups. Concerning mammals, it is well known that large forms evolved from small ancestors for almost every living and extinct Order. Many examples that come to mind, however, seem to deny the applicability of Cope's Rule. For instance, during the Pleistocene, many giant forms of familiar animals were known: giant deer, sloths, lemurs, sheep, baboons. These animals, however, are not the ancestors of their modern counterparts but members of evolutionary sidelines that, branching off from the main line of evolution, passed on to extinction. There are numerous examples of mammals to support the belief that an increase in body size, short of extreme giantism, is advantageous and is therefore subject to intense selection pressures.

Similarity of bodily form between two animals is most likely to result from a common ancestry, but there are many instances where unrelated or distantly related animals have come to look alike as a result of adapting in a similar fashion to equivalent environmental situations; this phenomenon is known as either convergence or parallelism.

The terms *convergence* and *parallelism* describe identical processes but, conventionally, they are employed in different contexts. Convergence implies that similar characters have developed in two stocks that are unrelated (i.e., in separate Orders), while parallelism indicates

31

similar adaptations in two stocks within the same Order.[5] The existence of both a placental and a marsupial rat, mole, rabbit, and dog are classic examples of convergence; while the evolution of an Old World and New World brachiating primate (gibbon and spider monkey) is one of the numerous examples of parallelism to be found amongst the primates. A brachiating mode of life, in fact, has evolved at least five times in five different primate families. This is to be expected when it is realized that brachiation is not really a particularly esoteric specialization, but is simply an expression of extreme arboreal adaptation. Tropical forest habitats, whether they are in Asia, Africa, South America, or Madagascar, whether they are in the Miocene or Recent, may differ floristically but not physionomically; their physical structure is much the same all the world over. Each zoogeographical realm has produced its own answer to a universal challenge.

---

[5] The distinction between the two terms is as much a matter of convenience as anything else. The author's usage is neither right nor wrong; it simply suggests a neat way to use the terms unequivocally.

CHAPTER THREE

The world of primates

THE MEANING OF CLASSIFICATION

The function of a classification is to arrange things or organisms in logical groups. The question of what constitutes a logical group of organisms is the art and science of systematic biology.

To Linnaeus, the father of modern systematics (1707-1778), it seemed logical to arrange animals in certain groups because they looked alike. But to other classifiers, it was more logical to arrange animals in accordance with similarities of habit or geographical realm. To the modern scientist, logical classifications are those that are based on genetic affinity. Genetic affinity may be represented by an ancestor-descendant relationship and classifications based on this principle are termed vertical or phylogenetic. Alternatively, genetic affinity may be expressed in terms of contemporaneous groups of animals of common ancestry, in which case the classification is termed horizontal. Simpson (1961) illustrates these two types of classification by analogy of the relationships between a man and his son (vertical) and a man and his brother (horizontal). Figure 3 is simply a synoptic list presented in the form of a diagram and indicates the relationships between contemporary groups of primates; Figure 4, on the other hand, is based on a geological time-scale and pro-

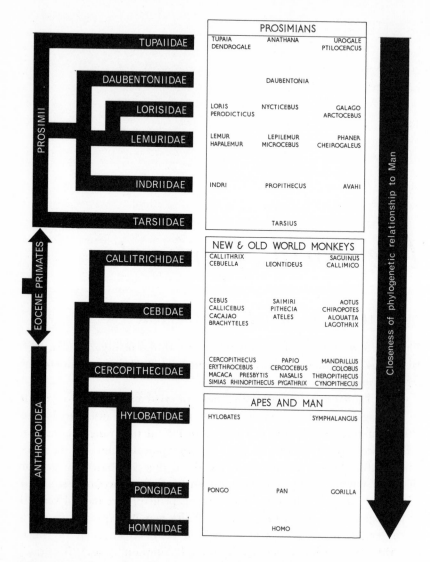

Figure 3. "Horizontal" classification of the primates illustrating relative closeness of living genera to man. (From "Prospects in Primate Biology," by J. R. Napier. *Proceedings of the United States National Museum*, 1968.)

vides in simplified form a picture of phylogenetic or evolutionary relationships between modern groups.

As has already been indicated in Chapter 2, resemblances need not necessarily result from inheritance from a common ancestor but can crop up in quite unrelated animals as a result of similar adaptive changes evolving under similar environmental conditions. This phenomenon is known as parallelism or convergence depending, simply, on the initial relationships of the two animals concerned (see p. 31). Resemblances between animals can also be derived from the operation of certain physical laws of growth which affect all animals irrespective of genetic relationship. The occurrence of bony crests on the skulls of large primates provides an example of the phenomenon called *allometry*. Large animals have large jaws, and consequently develop a bony crest running fore-and-aft on the top of the skull. This so-called sagittal crest is an adaptation which provides supplementary anchorage for the massive jaw musculature. Crests are found in, for instance, male gorillas, baboons, and colobus monkeys, but their presence does not necessarily indicate a particularly close genetic relationship between these animals; indeed saggital crests are found in all large-jawed animals of other Orders such as tigers, bears, and elephants. Resemblances, in fact, can be very misleading and the function of systematicists (or taxonomists) is to determine what is relevant or irrelevant for classificatory purposes.

Linnaeus' classification of animals was based on the archetype principle which took as its keystone a "typical" member of the group. It must be remembered that, in Linnaean philosophy, animals were expressions of special creation—a tiger was a tiger was a tiger; it was created as such and was destined to remain, immutably, in its original image. Departures from the archetype (variations in our sense) were regarded as "accidents" and of no taxonomic significance. Genetic variation of course, as we now appreciate, is the principal source of evolutionary change.

At first reading it may seem rather surprising that Linnaeus' classification differs, relatively speaking, so little from modern versions, considering how disparate are the basic premises. The reason that both classifications resemble each other so closely is that both are based primarily on similarity. Animals that *look* alike very often share a common ancestry but, as we have already said, this is not always so. Linnaeus, being unaware of the evolutionary theory, knew nothing of the principles of adaptation, parallelism, and convergence and was, therefore, quite incapable of distinguishing between animals that looked alike because they were (genetically) alike and animals that looked alike because they had grown (environmentally) alike over the course of time.

The system of codification by rank, the hierarchy, devised by Linnaeus, originally contained five categories: Kingdom, Class, Order, Genus, and Species. Since then new categories have been introduced and the taxonomic standing of old categories has become modified. Reference to standard works of systematics will reveal the whole extent of the hierarchical framework; here, only those taxa commonly used in primatology are shown:

Order
Suborder
Infraorder
Superfamily
Family
Subfamily
Tribe
Genus
Subgenus
Species
Subspecies

The synoptic list of primate genera shown in Figure 3 is arranged in a rather unfamiliar way. The arrangement was designed to emphasize the affinities of non-human primates to man. Thus, we have treeshrews (*Tupaia*) as man's most distant primate relatives and the African great apes, the gorilla (*Gorilla*) and the chimpanzee (*Pan*), as his closest. On the left of the diagram you will see the primary subdivision into two main groups or suborders, the Prosimii and the Anthropoidea. The constituent families of these suborders appear on the horizontal lines and their constituent genera at the appropriate level in the boxes on the right. Species are omitted from this diagram.

CHARACTERS OF PRIMATES

Prosimian primates differ from the Anthropoidea in a number of anatomical features, and it is in just these features that the Prosimii resemble non-primate mammals. Some lemurs, particularly the ring-tailed *Lemur catta* look, at a superficial glance, very like raccoons. It is interesting to

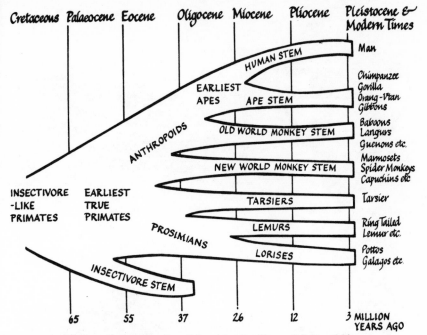

Figure 4. "Vertical" classification of the primates (From *Origins of Man*, by John Napier. Courtesy of Bodley Head, Ltd., 1968.)

analyze the basis of this similarity and, in the process, we may learn just what it is that makes primates different from all other mammals. (1) Raccoons and lemurs have much the same head and body size in terms of overall dimensions. Lemurs, however, are gracefully built and have disproportionately long hindlimbs; raccoons have rather chunky limbs of approximately equal length. (2) Raccoons have short banded tails. The lemur tail is also banded but is much longer than the length of the head and body combined. (3) The faces of both have long pointed snouts, moist noses, and tethered upper lips such as one sees, for instance, in dogs. Both animals possess patches of black fur around the eyes; in *L. catta* these take the shape of goggles and in the raccoon the black patch looks more like a bandit's mask. (4) The hands and feet of both species bear five digits, but there the similarity ends; the raccoon's digits bear claws while the lemur's fingers and toes are surmounted by flattish nails except for a single "toilet" or scratching claw on the second toe. (5) The innermost digit of the hands and feet of lemurs is separated from

37

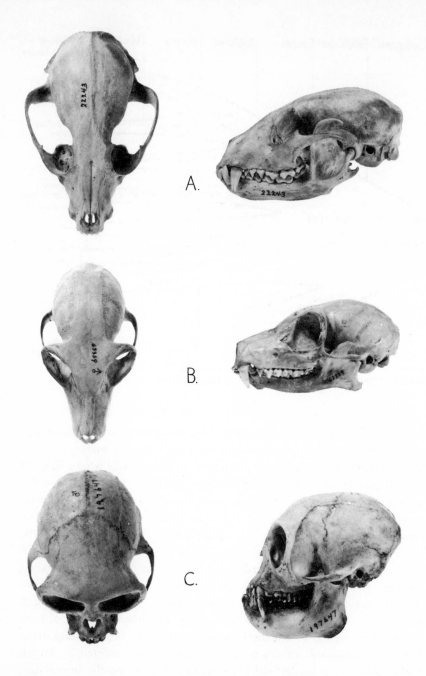

Plate 1. Dorsal and side views of skulls. (A) Raccoon *(Procyon)*.
(B) Lemur *(Lemur)*. (C) Colobine monkey *(Presbytis)*.

the rest and is opposable (see Chapter 8) and is capable of being moved freely and independently; there is no such distinction between the digits of the raccoon and they all have much the same functional value. (6) Raccoons have three pairs of nipples but lemurs usually have only one pair placed high up on the chest; however, occasionally an extra pair of nipples is found on the lower abdomen in lemurs (Schultz, 1948).

Some further interesting distinctions become apparent when we look at the skulls (Plate 1). (7) Raccoons (Plate 1A) have 20 teeth in each jaw whereas lemurs (Plate 1B) have 18. Mammalian teeth are of four functional kinds—incisors, canines, premolars, and molars. The dental formula[6] for the raccoon is $\frac{3.1.3.3.}{3.1.3.3.} = 40$ and that for the lemur is $\frac{2.1.3.3.}{2.1.3.3.} = 36$. (8) A ring of bone completely surrounds the eye-socket of lemurs, but is incomplete in the raccoon. The optical axes of the sockets themselves are directed forwards in the lemur and sideways in the raccoon. (9) The snout of the raccoon is proportionately longer than that of the lemur and the "scroll" bones, or turbinals, inside the nose are more numerous and complicated, correlated with a more highly developed sense of smell. (10) The circular opening at the base of the skull, the foramen magnum, transmits the spinal cord to the vertebral column. The position of the foramen at the base of the skull provides an indication of head and body posture. The foramen magnum of the raccoon lies well towards the back of the skull in a typical position for non-primate mammalian quadrupeds, while that of the lemur lies further forwards, reflecting the lemur's semi-upright sitting and resting posture (11). The braincase of the raccoon is smaller and flatter than that of the lemur.

Some of the important physical and behavioral characteristics of primates will emerge as we analyze these differences. Numbers in parentheses refer to the numbered distinctions noted above.

Primates are essentially arboreal animals whose limbs are adapted to leaping and climbing in the trees (1). Active arboreal locomotion demands the mechanical assistance of a long tail (2) and prehensile hands and feet with opposable thumbs and big toes to aid climbing and to ensure stability on slender branches high above the ground (4 and 5). Active arboreal locomotion also demands a much more accurate judgment

---

[6] The dental formula is customarily written in this way. The sequence of digits from left to right indicates the number of incisors, canines, premolars and molars in one half of the upper jaw (above) and the lower jaw (below). The digits on the right of the equals sign indicate the total number of teeth in the mouth.

of distances than life on the ground. This is brought about, in part, by the development of binocular vision which can only come into play if the fields of vision overlap one another. The more forwardly facing are the eyes, the more overlap of visual fields is present (8). Arboreal life does not require as highly developed a sense of smell as life on the ground; primates therefore show a reduced olfactory efficacy compared with other mammals. Noses are shorter, turbinal bones reduced in number and the olfactory bulb, the receptive center for smell in the brain, is diminished in size (9). The relative dominance of the visual sense in lemurs and the increased manipulative function of hands and feet lead to enlargement and elaboration of the brain (11). Habitat is clearly related to the nature of the diet and arboreal animals tend to be less omnivorous than ground-living forms; this habit is reflected among other things in the size and number of the teeth (7). A major trend of primate evolution, which will be discussed in more detail in Chapter 4, is towards bodily uprightness. This reflects itself in the bodily posture of primates, all of which are capable of sitting upright (10). Primates, as a rule, give birth to single offspring; except in certain primitive species such as tree-shrews, mouse lemurs, marmosets and tamarins, twinning is as rare as in humans.

Although the lemur does not show the characteristics of the primate order as comprehensively as does a monkey or an ape, it demonstrates all the principal physical *trends* that, taken together, identify a primate. Viewed singly, there are few characters that are unique to primates; many of them are shared with other mammalian Orders but, collectively and inimitably, they define this zoological group.

Structural and Behavioral Trends in Primate Evolution
(After Clark, 1959; Napier and Napier, 1967)

1. Preservation of a primitive mammalian structure of limbs, e.g., retention of the five-digit pattern of hands and feet, and of the clavicle, the radius and the fibula—bones that are reduced or absent in some groups of mammals.

2. A progressive freedom in mobility of the digits, especially the thumb and the big toe.

3. The replacement of sharp claws by flattened nails associated with the development of sensitive touch pads on the tips of the digits.

4. A progressive shortening of the snout.

5. An increase frontality of the eyes associated with the development of binocular vision.

6. Reduction in the apparatus and function of smell.

7. The reduction in number of teeth and the preservation of a simple molar cusp pattern.

8. Expansion and elaboration of the brain, particularly those regions concerned with vision, muscular coordination, tactile appreciation, memory and learning.

9. Progressive development of truncal uprightness.

10. Progressive elaboration of the placenta, with particular respect to the intimacy of the blood circulation between the mother and the fetus.

11. A greater dependency in locomotion on the forelimbs at the expense of the hindlimbs.

12. Prolongation of prenatal and postnatal life periods.

13. Increase in body size.

14. The development of a complex social system involving, progressively, a greater number of individuals whose hierarchal status must be taken into account.

PRIMATE FAMILIES

Following the sequence shown in Figure 3, a brief account of each living primate family concludes this chapter.

The Prosimians   The family consists of five genera of treeshrews presently recognized: *Tupaia, Ptilocercus, Anathana, Urogale,* and *Dendrogale.* They are found in India (*Anathana*), Southeast Asia (*Tupaia, Ptilocercus,* and *Dendrogale*), and in the Philippines (*Urogale*). They live both in the trees and among the low shrubs of the forest floor. With the exception of *Ptilocercus,* the feather-tailed treeshrew, which is nocturnal, treeshrews are diurnal. Generally speaking they are drab in color, somewhat rodent-like in appearance with long bodies and tails, prominent muzzles, moist noses, and tethered upper lips. Their hands and feet are clawed.

The patterns of social behavior are little understood at present, but there is some evidence to suggest that treeshrews form social groups with

41

male dominance hierarchies. The reproductive behavior of *Tupaia* is peculiar in that the infants are reared in a separate nest from the mother and are only fed once every two days (Martin, 1966). Failure in the past to appreciate this unusual maternal behavior has led to great difficulties in breeding these primates in captivity.[7]

The family Daubentoniidae (aye-ayes) contains a single genus and a single species. Their habitat is the tropical rain forest of Madagascar where they are now regarded as an endangered species. Efforts have been made towards conservation by placing several pairs on a small island off the coast of northwest Madagascar.

Aye-ayes are nocturnal animals that live on a diet of insects and fruit. They are cat-sized with long, coarse, and predominantly dark fur; the tail is long and bushy. As in all "lemurs," the hindlimbs are considerably longer than the forelimbs. A number of special peculiarities set the aye-ayes apart; the upper incisor teeth are continuously growing, long and rodent-like, with chisel edges; and the middle finger of the hand is fantastically slender, looking rather like a piece of bent wire. These two structural peculiarities are closely linked with the unusual feeding habit of aye-ayes. A popular item of the diet is the larvae of wood-boring beetles of the genus *Oryctes*. When the grubs are located by a combination of hearing and the sense of smell, the long incisors are used to gouge away the bark and outer wood layer over the flight holes; the wire-like finger is then introduced into the enlarged flight-hole and the grub is scooped out. The aye-aye is the only primate that possesses inguinal nipples only.

The family, Lorisidae (lorises), all members of which are nocturnal, includes two distinct subfamilies—Galaginae (the galagos) and Lorisinae (the true lorises).

The true lorises include four genera, *Perodicticus* (the potto), *Arctocebus* (the angwantibo), *Nycticebus* (the slow loris), and *Loris* (the slender loris). The first two genera are African and the last two are found in India, Ceylon, and Southeast Asia. All are arboreal and occupy tropical forest, monsoon forest, or mixed deciduous woodland. All four genera have strikingly large eyes facing forwards and small ears. As a group they are characterized by having very short or absent tails. Body proportions

---

[7] There are a number of authorities today who do not accept the treeshrews as primates. The evidence derived from anatomy is equivocal. Students who wish to know more about this problem are referred to R. D. Martin (1967). Treeshrews are still included among the primates in this book as the author, lacking personal experience, prefers to take refuge in the status quo.

are variable: pottos and slow lorises have rather chunky bodies with relatively short limbs; but slender lorises, as their name implies, have rather longer limbs of remarkable slimness. The Asian lorises have a white stripe down the nose separating the dark eye patches; *Loris* has a more pointed muzzle than that of *Nycticebus*. Of the African lorises, *Perodicticus* has a broader, blunter muzzle than that of *Arctocebus*; neither has any striking facial markings.

The hands and feet are remarkable in having huge, very specialized thumbs and big toes that are set way apart from the other digits. When the potto's hand closes on a branch, the thumb rests on one side and the fingers on the other to form a powerful clamp. It is difficult to distinguish hands and feet when they are seen bunched together, except by the presence of a long toilet-claw on the second digit of the foot. All genera show a tendency to shorten the index finger; pottos and angwantibos show the extreme form of this specialization where it is little more than a nubbin of flesh and bone.

The locomotor behavior of Lorisinae is characterized by the almost ludicrous slowness and cautiousness of their movements. Nothing is really known about their social behavior in the wild but they are generally thought to be solitary animals, as most nocturnal forms are. A behavioral pattern that seems to bind the true lorises together as a group is that all species 'park' their infants by hanging them from a branch like an umbrella while they themselves go foraging (Alan Walker—personal communication).

The subfamily Galaginae (galagos) or bushbabies come in several different sizes: the large cat-sized *Galago crassicaudatus*, the tiny *Galago senegalensis*, and the even smaller *Galago demidovii*. But size apart, they are a fairly homogeneous group. Galaginae are found only in Africa. They are nocturnal and, in contrast to the Lorisinae, are extremely active and fast moving. This behavioral difference is reflected in two special features which they do not share with the slow moving Lorisinae. The ears of Galaginae are large, dish-like affairs capable of being turned in all directions like radar saucers—and no doubt to some extent they act like them too. The hindlimbs of galagos are astonishingly long and are especially adapted for leaping. Vertical jumps of over seven feet have been recorded for *G. senegalensis*. Little is known about the social behavior of galagos in the wild but they do not appear to be strictly solitary; sleeping nests of *G. senegalensis* may contain up to nine individuals. Infants, however, are raised by the mother in strict solitude.

43

Members of the family Lemuridae (lemurs) are found only in Madagascar. The lemur family, like the loris family is made up of two subdivisions (subfamilies) which, while basically similar in bodily and dental structure, show physiological and behavioral differences that are clear-cut enough to make it easier to describe them separately.

There are three genera in the subfamily Lemurinae (true lemurs): *Lemur, Hapalemur,* and *Lepilemur.* The first two are diurnal or crepuscular (active in the few hours around dawn and dusk) while *Lepilemur* is wholly nocturnal. All are arboreal forms living in the moist evergreen tropical forest of Madagascar's east coast, or in the dry deciduous tropical forest of the west coast. One species, *Lemur catta,* is the "baboon" among the lemurs; it is partly ground-living and occupies the dry thickets and woodland savanna country of southwestern Madagascar (Plate 2). It is instructive to note that of all the lemurs *L. catta* is the most successful in captivity; like the ground-living macaques and baboons, it is relatively tough and resilient, able to live under a variety of ecological conditions and utilize a wide variety of foods. Ring-tailed lemurs have adopted a highly organized social life on the baboon troop principle of a multi-male society run by dominant males. This instance of parallelism between prosimians and anthropoids indicates the great influence of the environment in steering the course of evolution. All three genera of Lemurinae can be classified in locomotor terms as vertical clingers and leapers (see Chapter 5).

In appearance, the Lemurinae differ quite considerably but this is largely due to variations of coat color, a trait that has run riot among the genus *Lemur.* In one species, *L. variegatus,* three color varieties have been described varying from white with some black, and black with white transverse body bands to red with black bands. The species *L. macaco* shows sexual dichromatism, the male being all black and the female reddish brown.

In size Lemurinae vary between that of a cat (*Lepilemur, Hapalemur,* and *L. catta*) and a cocker-spaniel (*L. variegatus*).

As in all Madagascan lemurs, African and Asian lorises, and galagos, the hands and feet are prehensile by virtue of the structural separation of the thumb and the big toe. The teeth of the true lemurs are 36 in number with a dental formula of $\frac{2.1.3.3.}{2.1.3.3.}$. The upper incisors, however, are extremely small in all genera; *Lepilemur* lacks these teeth completely. The arrangement of the incisors and canines of the lower jaw is very strange. As in all Lemuridae and Lorisidae, the incisors and canines form

Plate 2. Ring-tailed lemurs *(Lemur catta)* at Chester Zoo, England. (Photograph by Doris Sorby.)

a series of procumbent, closely packed, slender teeth which, together, resemble nothing so much as the tines of a comb. In fact it is usually called a "dental comb" and its function is largely cosmetic. Instead of combing the hair of the head, however, it is used to clean and arrange the hair of the body.

The true lemurs depend very largely on the sense of smell as the principal channel of social communication. They mark out their territory (very much as a dog does with urine), but using the secretion (pheromones) of certain cutaneous glands to convey information to other lemurs. To facilitate this trail-laying and territory-marking behavior, male lemurs are equipped with special scent glands on the inner aspect of their forearms just above the wrist and at the junction of the arm and the shoulder. Lemurs can often be observed anointing their tails with their forearm glands and then disseminating the scent by waving and flicking

45

the tail which acts as a sort of built-in censer. Another method of territorial marking, often seen in captivity, is the gesture of rubbing the perineal region against the wall, the bars or upright furnishings in the cage. The perineal region is richly equipped with scent glands. Apes and man have only a few scent glands left. In man they occur in his armpit and in his groin, but they are seldom consciously used as means of communication. Although their exploitation is not part of westernized culture, they have a certain sexual significance to mankind as a whole.

The diet of Lemurinae is uniformly vegetarian but varies in the details of the proportion of fruits and leaves that various species consume. The social behavior of Lemurinae is not well known for all species, but generally speaking they are social animals which live in troops; only *Lepilemur,* the nocturnal form, is truly solitary.

The subfamily Cheirogalinae also consists of three genera: *Cheirogaleus* (dwarf lemurs), *Microcebus* (mouse lemurs), and *Phaner* (fork-marked dwarf lemurs). All are arboreal and nocturnal; all rest in holes in trees and in hollow bamboos, and all are insect-eating. *Microcebus,* the smallest of the Cheirogalinae, is also the smallest of all primates. Its head and body length is 130 mm and its weight 40 grams; but, in spite of this, the brain of *Microcebus* shows a number of advanced characters which place this tiny dwarf into an intelligence category well above that of many much larger non-primate mammals.

The Cheirogalinae are undistinguished in color. *Phaner,* in justification of its common name "fork-marked," presents a dark, well-defined, spinal stripe which splits into two on the crown of the head, each segment of which becomes continuous with the dark rings around the eyes. This subfamily has an interesting sexual cycle. The females are only fertile at certain seasons of the year. During nonfertile seasons the vaginal orifice is said to be completely closed over, sealed against all invaders by furbearing skin (Petter-Rousseaux, 1964) ; this extraordinary version of a medieval chastity belt has also been observed in galagos (Butler, 1967).

Socially, Cheirogalinae are generally believed to be solitary animals. In captivity, however, they seem to be able to live gregariously.

The family Indriidae (indrises, sifakas, and avahis) is perhaps the most remarkable of all Madagascan primates and is possibly one of the most ancient. Indriids are animals of moderate-to-large size and of striking appearance. They share a very primitive mode of locomotion for which the principal adaptations are extremely long legs, a vast, grasping big toe, and an upright resting posture (see Chapter 5). Indrises and

46

sifakas are diurnal forms and are strikingly marked in black and white, sometimes with reddish patches. The indris (*Indri*), which lacks a tail, has a black face, back, thighs, and extremities while the sifaka (*Propithecus*), with a *long* tail, is distinguished by having the black face and crown separated by a prominent white band. This white forehead band is also seen in the smaller, woolly lemur (*Avahi*), a nocturnal form which

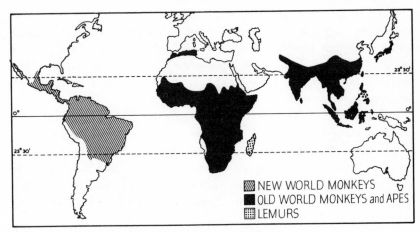

Figure 5.   Geographical distribution of living primates.

is otherwise a sombre grey-brown. The dental formula for the Indriidae is $\frac{2.1.2.3.}{2.0.2.3.} = 30$. Compared with the Lemuridae, the number of premolars is reduced, and the "dental comb" is formed by the four lower incisors only, the canine being absent. Indriidae prefer a vegetarian diet with a high cellulose content derived largely from leaves.

Socially, Indriidae are highly territorial, defending their territories by a vocal form of "formalized" aggression (cf. howler monkeys and gibbons). All the Indriidae, as far as is known, live, like gibbons, in family groups.

The family Tarsiidae (tarsiers) contains only one living genus, *Tarsius*; the three recognized species are restricted to some of the islands of Malaysia, Indonesia, and the Philippines.

Tarsiers are amongst the smallest of primates, the top weight being little more than six ounces (165 grams). They are tropical rain forest forms that exist almost wholly on live animal protein, i.e., insects, lizards and spiders. Structurally they are characterized by enormous, forwardly facing eyes which betray their nocturnal habit; short forelimbs and long,

47

slender frog-like hindlimbs make tarsiers prime exponents of the loco-motor pattern called vertical clinging and leaping (see Chapter 5).

The Anthropoids The anthropoid primates comprise the Old and New World monkeys, the apes, and man. They represent a distinct advance on the prosimians both in terms of structure and behavior—of particular structural note are the eyes, the muzzle, the brain, and the hands.

While the *eyes* of prosimians are considerably more frontally rotated than those of raccoons, the eyes of monkeys have completed the process and look directly forwards (Plate 1C). This position of the eyes provides for the overlap of visual fields, the necessary basis for stereoscopic vision or vision in depth. The bony eye-sockets of monkeys are fully closed off, while in prosimians they are still open at the side.

The *muzzle* of most anthropoids is fairly short, unlike the projecting snout of many of the prosimians (Plate 1). This reduction is associated with a diminished emphasis on the sense of smell. To a large extent increased visual acuity has replaced olfactory acuity. Some monkeys, however, such as baboons have secondarily evolved a long muzzle; this adaptation is related not so much to a more active sense of smell as to the mechanical requirements of large powerful jaws.

The *brain* is considerably larger in anthropoids than in the prosimians. For instance, the brain of a marmoset is three times bigger than that of a galago in spite of the fact that the animals are of comparable size. Bigger animals have absolutely bigger brains, so mere size can be rather misleading as an indicator of increased intelligence. But, if an adjustment is made so that the factor of size is ruled out, then we see a truer picture of the evolutionary improvement in the brain as we ascend the scale from monkeys to apes and from apes to man (Figure 16). Size, however, is not the only criterion by which increased intelligence can be judged. The surface of the cerebral hemispheres, which are the parts of the brain that receive incoming sensory impulses and initiate voluntary movement, become more and more folded as the underlying brain substance expands. Expansion is due largely to an increase in the number of the fibers connecting the brain with the spinal cord and one part of the brain with another. The cerebral cortex of prosimians shows very few fissures but that of the monkey shows considerably more; by the time the ape level is reached, the fissures and folds are almost as complicated as they are in man. The greater brain complexity of the anthropoids over the prosimians is reflected in the behavioral differences apparent between

48

these two suborders. Anthropoids, generally speaking, live more complicated social lives than prosimians; they have richer and more varied means of communication and they are more capable of learning new patterns of behavior. These advances, theoretically, should provide them with a greater capacity for evolutionary survival in the face of competition with other species in times of great environmental change. The lemurs that we know today are a tiny remnant of the great lemur empire of the Eocene era. The remaining few owe their survival to the fact that they live in "refuge" areas where there is little competition from higher primates or any other kind of competing mammal.

Lacking geographic isolation, certain prosimians have survived in Africa and the Far East by adopting a nocturnal way of life which effectively removes them from competition with the higher primates which in these areas are exclusively diurnal.

One of the reasons that anthropoids have more advanced brains than prosimians is that their *hands* are capable of a greater variety of complex activities. The question of the evolution of the primate hand is discussed in Chapter 8. Suffice it to say here that the basis of increased manual activity lies largely in the development of an *opposable* thumb.

Apart from these four important distinctions, there are differences in the *teeth,* in the *limbs* where the prosimian to anthropoid trend has been to increase the length of the forelimbs relative to the hindlimbs, and in the form of the *placenta* which differs profoundly from that of prosimians in respect of the intimacy of the fetal and maternal blood circulations. The evolution of the placenta in the primates is of great importance to the pattern of evolution of the Order as a whole, but it is a complex subject which we cannot discuss here.

New World Monkey Families As a group, the Neotropical primates are distinguished from Eurasian and Ethiopian forms by a constellation of characters, none of which are particularly striking when taken in isolation. In the dentition, three premolars are retained in each half of each jaw. The skull is distinguished by the presence on its undersurface of a bony balloon called the tympanic bulla, which represents the expanded wall of the cavity of the middle ear. The external portion of the ear comprises a longish bony tube in Old World monkeys, but in New World forms, the tube is extremely short, little more than a ring in fact. Perhaps the most obvious distinction between living Old and New World monkeys is in the arrangement of the nostrils. In New World monkeys the nostrils face sideways and are far apart, separated by a broad nasal septum; in Old World forms they face downwards or forwards and are close together (Figure 6). This obvious distinction has given rise to the alternative names for these two groups which will be frequently employed in this book: platyrrhines (broad-nosed monkeys) and catarrhines (narrow-nosed monkeys). The thumb of platyrrhines is *non-opposable* in the special sense that this term is used (see Chapter 8). The big toe is fully opposable as in all non-human primates with the exception of the treeshrews.

Callitrichidae (marmosets and tamarins) is a geographically widespread family of New World monkeys which consists of five genera—*Callithrix, Cebuella, Saguinus, Leontideus,* and *Callimico*—and 17 species. They are squirrel-sized animals. The smallest of the family, *Cebuella pygmaea*, the pygmy marmoset, has a head and body length of 130-145 mm. The largest member of the family is *Leontideus rosalia* (the golden lion marmoset) whose body length varies from 227-370 mm. The difference in size between sexes of Callitrichidae is not marked; in fact females are sometimes larger than males. Fur color is extremely varied and there are many striking color patterns; facial adornments such as moustaches and ear tufts are a prominant characteristic of Callitrichidae.

The marmosets and tamarins are distinguished from other New World monkeys by their small size and the possession of claws instead of nails on all digits (except the big toe). The dental formula: $\frac{2.1.3.2.}{2.1.3.2.} = 32$, includes only two molar teeth in each half of each jaw; Callitrichidae are unlike the Cebidae which universally have 3 molars. *Callimico* (Goeldi's marmoset) is an exceptional genus; it has claws like other marmosets but 36 teeth like all the Cebidae (Plate 3).

Little is known of the social behavior of Callitrichidae in the wild, but they are assumed to live monogamous lives in family groups. The male parent carries the young except when the mother is nursing. Twin births are the rule.

The Cebidae (New World monkeys other than marmosets) is a diverse family which includes 11 genera and 29 species; it is divided into five subfamilies: Atelinae (the spider monkeys and woolly monkeys), Alouattinae (the howler monkeys), Aotinae (the night monkeys and titis), Cebinae (the capuchins and squirrel monkeys), and the Pitheciinae (the sakis and uakaris).

The Atelinae show the most advanced specializations of all cebids. Effectively they possess five, not four, limbs. This needs a bit of explaining.

In this subfamily the tail is prehensile; that is to say, it is capable of grasping objects just as efficiently as is the hand or foot. On the under-surface of its terminal third there is an area of naked skin. On microscopic examination, this skin has the histological characters of finger skin—the corrugations or "finger-prints," the sweat glands, and the microscopic nerve endings called Meissner's corpuscles that serve the sense of touch. Spider monkeys often use the tail instead of the hand for grasping, even small objects like raisins or peanuts can be picked up by the tail. Apart from its grasping function, the prehensile tail has an important function in locomotion. Atelinae are basically quadrupedal monkeys but they fre-quently progress through the trees by means of arm-swinging; during this activity, the tail is normally used as an extra limb. It is common sight to see spiders and woollys hanging from an overhead support by means of the tail alone.

Spider monkeys can be distinguished from woolly monkeys by their much longer limbs (and higher intermembral index, see Table 8), by the relative shagginess of their coats (woolly monkeys have denser, plush-ier fur), and by the absence of a thumb. This thumbless state is almost unique amongst primates; the only other primate to show this peculiar adaptation is the catarrhine monkey, the African genus *Colobus*.

The subfamily Alouattinae is composed of one genus *Alouatta* (the howler monkeys). Howlers are large, robust, prehensile-tailed monkeys with longish limbs. They are predominantly leaf-eaters. They vary in color, according to species, from black (*A. caraya*), brown or buff (*A. fusca*) to copper-red (*A. seniculus*). The peculiar specialization of howlers relates, as you might guess, to their voice. In its penetrating quality, it is only matched by the gibbons'. Howling sessions between different groups are a regular early morning occurrence and are believed to serve the func-tion of defining and "defending" living space, a substitute, if you like, for physical aggression. The anatomical basis for this howling behavior is in the structure of the hyoid bone which, in man and other primates, is a slender, horseshoe-shaped affair. The howler hyoid has become expanded into a gigantic egg-shaped bony box which forms a resonating chamber for the voice. This large structure under the chin has had a considerable effect on the shape of the howler's jaw and skull. As a consequence, these animals give the impression of having both an oversized head and an oversized inferiority complex, making it necessary for them to hunch their shoulders and bury their heads on their chests.

Another subfamily of the Cebidae, the Aotinae, contains two genera,

51

Aotus (the night monkey) and *Callicebus* (the titis). The night monkey or douroucouli is the only nocturnal monkey in either hemisphere; they are small-to-medium-sized primates with thick, soft, brown or greyish coats. On the head there are usually three longitudinal dark stripes and, over each eye, a semilunar patch of white fur that, when the eyes are closed in sleep, gives the disconcerting impression of a pair of wide-awake eyes staring straight ahead. Night monkeys are thought to have a similar social organization to that of marmosets and titis, i.e., small family groups of 2 to 5 animals; also like the titis and marmosets, male night-monkeys often carry their infants on their backs. This practice is seen only in the most primitive of the New World monkeys and has never been recorded in Old World monkeys or prosimians.

The titis are about twice the size of marmosets, their fur is long and thick, and the tail is bushy and nonprehensile and is much used during locomotion as a balancing organ. In their skull anatomy and their vocalization behavior, titis are rather like miniature versions of the much larger and more specialized howler monkeys.

The Cebinae also contains two genera: *Cebus* (the capuchins) and *Saimiri* (the squirrel monkeys). Squirrel monkeys are small, lithely built animals with short fur and neat, oval-shaped heads; the limbs and underparts are yellowish and the back grey-green. Capuchins are much more robust, weighing nearly three times as much as squirrel monkeys. The predominant color of the fur varies according to species, from dark brown in *Cebus apella* to black and white in *C. capucinus*; all species show a dark-colored crown patch. The capuchin's tail is semiprehensile and is often curled around a branch when at rest. Capuchins are alert, highly intelligent animals whose behavior, when it is studied in the wild, should be very illuminating. Only preliminary field studies have so far been made (Thorington, 1967). In captivity, they are ingenious, quick to learn, manually efficient, inexorably curious and very aggressive—rather reminiscent of humans in fact.

The Pitheciinae are a subfamily containing three genera: *Cacajao* (the uakari), *Chiropotes* (the bearded saki), and *Pithecia* (the saki). They are small-to-medium-sized animals, weighing approximately 3 kilograms and with a body length of 45 centimeters. The uakari is perhaps the strangest looking of all primates. One species, *Cacajao rubicundus*, has a bright crimson face, a shaggy dark red coat and it moves with a curious gait with the hands turned out sideways. Uakaris traditionally have been

Plate 3.   Goeldi's marmoset *(Callimico goeldii)*. Male carrying 8 weeks of infant. (Courtesy of Rainer Lorenz.)

maligned as sluggish animals but this is very far from the truth; on the contrary they are extremely energetic. The Pitheciinae are characterized by a most remarkable arrangement of the incisor teeth; both upper and lower incisors are procumbent so that the "bite," instead of being vertical, is nearly horizontal. Nothing is known about the functional significance of this most curious dental arrangement, which resembles the "dental comb" of the lemur, but it is presumably related to dietary behavior.

In addition to the morphological characters of platyrrhines common to all Callitrichidae and Cebidae mentioned on page 49, certain ecological and behavioral characters of the superfamily as a whole can now be added. All platyrrhine monkeys are arboreal; no ground-living forms exist. They are quadrupedal in locomotion. Cebid diet consists substantially of fruit and leaves, but for the smaller members of the platyrrhine families such as marmosets, tamarins, titis, and night monkeys, insects form a major part of the diet. Generally speaking, platyrrhine monkeys live in large bands with a multi-male society structure; the exceptions are the insectivorous genera mentioned above, which are found in family units. It is not possible to be more precise because our knowledge of the social structure of platyrrhines (with the notable exception of the howler) lags far behind, both in quality and quantity, our knowledge of catarrhine behavior.

Although the platyrrhines *as a whole* are a more primitive group than the catarrhines, it should be borne in mind that they show many highly specialized and progressive characters which may be a reflection of a long period of optimal environments that they have enjoyed since their evolution in the Eocene.

Old World Monkey Families   Some of the principal characters shared by platyrrhine and catarrhine primates have already been referred to on p. 48. Now it is necessary to discuss, briefly, the characters which distinguish catarrhine and platyrrhine monkeys from each other. In fact these two geographically distinct branches of the suborder Anthropoidea are very hard to tell apart. There are certain dramatic characters such as the prehensile tail in platyrrhines that immediately differentiate the two groups. No Old World primate has a prehensile tail; but then neither have quite a number of New World forms, so this does not really constitute an absolute distinction. Perhaps the most useful guide, already mentioned, is in the characters of the nostrils: wide open and wide apart

Figure 6. Comparison of the external nose in (A) catarrhine monkey and (B) platyrrhine monkey. (Courtesy of British Museum (Natural History).)

= platyrrhine condition; narrow and close together = catarrhine condition (Figure 6). Other structural distinctions between platyrrhines and catarrhines are discussed below.

The catarrhine hand is somewhat more dexterous, particularly in respect of the thumb, which is truly opposable; in platyrrhines it is only pseudo-opposable (see Chapter 8). Opposability provides an improved basis for manual dexterity. Behaviorally, Old World monkeys rely rather more heavily on their hands for various activities, such as feeding, grooming, and manipulation of objects within their environment, than do New World forms.

Locomotor activities show a much greater range of variation in catarrhine than in platyrrhine primates (see Chapter 5)—and the significant characteristic of upright sitting (significant, that is, for the emergence of bipedal man) is more highly evolved in the former. This behavioral activity has left a very positive mark on all catarrhine monkeys. If you turn them upside down you will find on their base not one, but two, seals of approval for this way of life—the *ischial callosities*. These are hard, cornified pads which, no doubt, ameliorate the discomfort of upright sleeping, perched on the hard, rough branch of a tree. No New World monkey is so blessed; their habitual sleeping posture is either in a hunched cat-like position resting on all four limbs or curled up on their sides. Ischial callosities are also present in gibbons and, in a modified form, in all great apes. Only man lacks these endorsements of catarrhine status; he has something much more useful—buttocks.

Cheek pouches are another catarrhine specialty, although not a universal one. These structures form capacious extensions of the cheeks that extend below the lower jaw almost meeting in the midline. Pouches can be crammed with food which can, at leisure, be pushed back into the mouth, usually by hand, and chewed and swallowed in the normal way. These useful accessories, reminiscent of the bulging cheeks of hamsters, are developed, however, only in the cercopithecine subfamily of the Cercopithecidae. The adaptive significance of cheek pouches is not at all easy to understand. As they are absent in the arboreal colobine subfamily, the inference is that cheek pouches are in some way connected with a ground-living way of life. As an adaptation for quick raids on native plantations, they are ideal 'shopping-baskets'; but clearly cheek pouches antedate African agriculture so their survival value cannot be quite so easily explained away. There are, however, dangers on the ground other than irate African Mr. MacGregors, and it was possibly the prevalence of ground-living carnivores that initially prompted the selection of cheek pouches as a feeding adaptation.

Sexual swelling is another phenomenon which is unique to Old World monkeys. Here again, it does not occur universally. During estrus, a phase of the menstrual cycle which accompanies the shedding of the ovum from the ovary into the uterine tube, the female external genitalia and the surrounding bare skin may become grossly swollen, protuberant, and highly colored. Periodic sexual swelling is characteristic of the macaques, baboons, and mangabeys, but even within these genera there is considerable variability in the occurrence of swelling and coloration. With the exception of the little talapoin monkey and Allen's swamp monkey, no other members of the species of the genus *Cercopithecus* (guenons) show this phenomenon, nor do the langurs or colobus monkeys.

Distinctions between Old and New World primates based on the structure of their social life are not particularly easy to make. Generally speaking, the most advanced and sophisticated systems within the primate order occur amongst catarrhines. This may be the result of the much greater ecological diversity of this group. No equivalent, for instance, exists among platyrrhines of the "harem" type of social structure of the gelada and hamadryas baboons of Ethiopia and the Sudan; but then no equivalent ecological situation exists (or has ever existed) in South America. Where similar ecological situations do exist, we find comparable forms of primate societies. Pair-bonding is characteristic of the marmosets and tamarins of the New World; it is also found amongst the gibbons of the Old World. Large bands composed of multi-male units occur amongst forest dwelling, fruit-and-leaf eaters in both hemispheres, notably the howler monkeys in South America and the colobus monkeys and langurs in Africa and Southeast Asia. With the present state of our knowledge of free-ranging primates, it is not possible to draw any hard and fast conclusions about social behavior that can provide a real basis for distinction between the two major groups. This lack of distinctiveness is further evidence that the social behavior of primates is somewhat of an *ad hoc* affair deriving its patterns largely from the exigencies of the environment, rather than from the dowry of the genes.

The vast family Cercopithecidae contains 14 recognized genera and 73 species. All catarrhine monkeys are grouped into one family containing two subfamilies, the Colobinae and the Cercopithecinae.

The subfamily Colobinae comprises the following genera: *Colobus* (the colobus monkey of Africa), *Presbytis* (the langur of India and Southeast Asia), *Nasalis* (the proboscis monkey of Borneo), *Simias* (Pagai Island, or pig-tailed, langur), *Rhinopithecus* (Snub-nosed langur of China) and *Pygathrix* (the Douc langur of Vietnam and Hainan). Colobine monkeys are characterized principally by their dietary speciality of leaf-eating. Leaf-eating is not an exclusive, but is probably a dominant, preference. This food habit, which is probably a very ancient one indeed, is associated with a number of important adaptations that relate to anatomy, ecology, and social structure.

All colobine monkeys are arboreal and, generally speaking, more acrobatic in trees than the cercopithecines. Certain species such as *Presbytis entellus* (the common Indian langur) are partially ground-living, an adaptation to the particular ecosystem that now operates in the agricultural and suburban areas of India where the forest cover has been largely destroyed. The arboreal adaptations of colobines are reflected in their limb proportions. Their hindlimbs are considerably longer than their forelimbs; this proportional relationship can be expressed quantitatively by the intermembral index (see Chapter 5). A high degree of arboreality is also indicated by the length of the tail which is considerably longer than the head and body length in all colobine species.

Anatomical and physiological adaptations to a diet of leaves affects the form of the teeth and the structure of the stomach, cecum, and colon, as well as the size of the salivary glands around the mouth. The stomach of colobines is sacculated (recalling, but not duplicating, the double stomachs of ruminants), the cecum and the colon are voluminous (correlated with the pot-bellied appearance of many colobines and of other bulk food eaters, such as gorillas and spider monkeys), and the salivary glands are particularly large.

The form of colobine societies, as far as we know them, consists of multi-male units with minimum male-dominance activity. The exceptions are *P. entellus*, the ground-adapted langur, a species which demonstrates sexual dimorphism, a characteristic of other male-dominated, ground-living primate societies such as the baboons.

Colobines go in for special noses. Why this should be so is quite incomprehensible. The nose of *Colobus* is prominent and overhangs the upper lip; that of *Nasalis* is very protuberant; in females it is sharp and up-turned; in adult males it is quite fantastically inflated and pendulous. The noses of *Rhinopithecus* and *Simias* are snubbed, and that of *Pygathrix* is characterized by grotesque flaps that are unmatched in any other primate species.

The subfamily Cercopithecinae is probably an offshoot of the primitive catarrhine monkey (possibly colobine) stock. The circumstances which produced the dichotomy of these two subfamilies was ecological, closely connected to the climatic and geomorphological events which were reshaping the world in the late Oligocene and early Miocene epochs (see Chapter 4).

The cercopithecines comprise the following genera: *Cercocebus* (the mangabeys), *Cercopithecus* (the guenons), *Cynopithecus* (the Celebes "ape"),[8] *Erythrocebus* (the patas monkey), *Macaca* (the macaques), *Mandrillus* (the mandrill and the drill), *Papio* (the baboons) and *Theropithecus* (the gelada baboon). One characteristic that these genera have in common is a propensity for living on the ground. As a group, Cercopithecinae present a heterogeneous bunch of (presently) arboreal forms—e.g., *Cercopithecus* and *Cercocebus*—and ground-living forms—e.g., *Cynopithecus, Erythrocebus, Mandrillus, Papio,* and *Theropithecus*—and an intermediate genus, *Macaca,* which is both arboreal and ground-living according to species. My belief, however, is that they all stem from a distant ground-living species of the Miocene that would have been more like *Macaca* than anything else.

Cercopithecines are characterized principally by their overall bodily structure which is more compact and less rangy than colobines. They have

---

[8] So-called because of the absence of a tail. The Gibraltar "ape" *(Macaca sylvanus)* is blessed with a similarly misleading sobriquet. Both animals are, of course, monkeys.

Plate 4.   Group of chacma baboons *(Papio ursinus).*
(Courtesy of Stephen Peet.)

forelimbs which are nearly equal in length to their hindlimbs, tails which are shorter, and muzzles which are longer than those of colobines.

Cercopithecines show a sexual swelling during estrus to a varying degree; they have cheek pouches; and they have longer thumbs and shorter fingers than colobines. In most genera, sexual dimorphism is strongly developed. In *Papio, Erythrocebus,* and *Theropithecus,* for instance, the adult male is almost twice the size of the adult female. Sexual dimorphism is also displayed in the size of the canine teeth which are universally longer in males. Males often show a conspicuous shoulder mane *(Papio* and *Theropithecus),* or a vividly colored mask *(Mandrillus),* or psychedelic genitals *(Mandrillus, Cercopithecus,* and *Erythrocebus).*

Social behavior in cercopithecines appears to be linked closely to habitat. Arboreal members of genera such as *Cercopithecus* and *Cercocebus* show patterns of social structure differing from those of the more terrestrial members of the same genera. The white-collared mangabey *(Cercocebus torquatus)* is largely a ground-living species and its social

59

structure broadly resembles that of baboons. The arboreal species, *C. albigena,* on the other hand, form groups of relatively few individuals with only a single adult male in attendance. Guenons, whose behavior in the wild is only known for a few species, show a similar pattern. The social unit of forest-living, arboreal guenons comprises small subgroups with single males although many such groups may be aggregated into much larger bands or parties. The most ground-living of the guenons (*Cercopithecus aethiops,* the savanna monkey) has social units with up to 40 individuals including many adult males. The presence of more than one adult male in a troop would lead to serious intragroup fighting, were it not for the stabilizing effect of a male-dominance hierarchy. Such a system demands certain behavioral patterns from the members of the troop, such as irritability towards inferiors and subservient acquiescence towards superiors; its principal advantage lies in the survival value for the group as a whole. At the top of the male hierarchical ladder are the alpha-males or leaders who wield, in different spheres, something amounting to absolute power. Absolute power may corrupt absolutely in a human context, but it spells survival in catarrhine primate societies, such as those of the baboons and the macaques.

Old World Ape Families   The apes are divided into two families, Hylobatidae (gibbons and siamangs), and Pongidae (chimpanzees, gorillas, and orang-utans). These families have certain features in common which distinguish them both from the catarrhine monkeys. The most obvious physical distinction is that monkeys have tails but apes do not. This might be classed as a Rule but not a Law, since there are certain monkeys that are virtually tailless. The Celebes "ape" and the Barbary "ape" are examples of tailless monkeys and the sobriquet has become attached to them for this very reason. Monkeys like the Moor macaque, the Japanese macaque, and the stump-tailed macaque possess the merest apology for a tail. Most monkeys, however, possess tails which are usually longer than their head and body length combined (see Chapter 6 for a discussion of tails).

Apes differ from monkeys in respect of their teeth, although at first sight the differences are not great. Both apes and monkeys possess 32 teeth and the same number of each kind, i.e., 2 incisors, 1 canine, 2 premolars, and 3 molars. The dental distinction between apes and monkeys lies in the anatomy of their cusps which reflect the differing adaptations of the two superfamilies with respect to diet. Apes are predominantly (but not wholly) fruit-eaters while monkeys are more omnivorous, eating leaves, fruits, grasses, roots and even an occasional small bird or mammal. The fruit-eating adaptations of apes are seen in the very broad spatula-like upper incisors and in the form of the molars which bear separate, rather conical cusps. Monkey incisors are narrower and the molar cusps are united transversely by crests (bilophodont condition). When the upper and lower teeth meet, the crests and valleys interlock like the serrated jaws of a wrench. This provides an ideal grinding surface for coarse vegetable foods such as leaves, roots, and grass. Ape molars, on the other hand, have somewhat sharp, conical cusps on their molars which are discrete, being separated by a characteristic arrangement of intervening grooves; in the lower jaw a fifth cusp has made its appearance. This cusp has never appeared in the first two lower molars of catarrhine monkeys but has developed on the last molar in most genera with the exception of *Cercopithecus* and the closely allied genus *Erythrocebus*. The separate cusp arrangement of apes' teeth more nearly approximates the primitive pattern of the primate dentition than the cross-linked cusps of monkeys. Thus, apes are more primitive than monkeys in respect of their teeth but, in respect of other systems such as the brain, the limbs, and the prolongation of life periods, the monkeys are more primitive than the apes. Further reference to these differences are more appropriately made in Chapter 6.

The family Hylobatidae contains two genera and seven species. The two genera are *Hylobates* (the gibbons) and *Symphalangus* (the siamang).

Gibbons and siamangs are arboreal, largely fruit-eating forms, whose special way of locomotion by arm-swinging has been termed brachiation (see Chapter 5).

Gibbons are slender, stylish-looking animals with exceptionally long arms and hands and relatively short legs. Their intermembral index is high (Table 8). Siamangs are larger and heavier than gibbons and have even longer arms. The skeleton and soft parts of the gibbon's body are modified in many ways for the essentially upright posture that it adopts during brachiation. Even when on the ground, their bodies are still upright as they walk with bent hips and knees, their hands seldom touching the ground. Both gibbons and siamangs have loud, prolonged calls which rise in inflection and pitch and increase in tempo and intensity. These calls are most often heard in the early morning and are regarded as having a territorial function, rather like that of the howler monkeys of South America. The siamang introduces a resounding and monumental boom in between each loud note of its call. This is produced by forcing exhaled

Plate 5. Young chimpanzees *(Pan troglodytes)* at Chester Zoo, England.
(Photograph by Doris Sorby.)

air into a curious sac-like compartment under its chin. This sac, which acts as a resonating chamber, blows up like a balloon as the air is forced in.

The social behavior of gibbons is based on the family group. Gibbons are monogamous and bear single young which are born at two-year intervals. As gibbons are slow to reach maturity, the family may consist of the two adults and one to four offspring.

The family Pongidae comprises three genera, *Pan* (the chimpanzee), *Gorilla* (the gorilla), and *Pongo* (the orang-utan); there are two species of *Pan* but only one of each of the others. The physical and behavioral characteristics of chimps and gorillas will be discussed fully in Chapter 6 so reference will only be made here to the orang-utan.

Orangs, inhabitants of Borneo and Sumatra, are extremely heavily

Plate 6. Male (right) and female (left) mountain gorillas *(Gorilla gorilla beringei)* at Chester Zoo, England. (Photograph by Doris Sorby.)

built animals, the males weighing 140–150 pounds and the females about 80 pounds; there is thus marked sexual dimorphism with regard to weight. Males and females differ in other ways; adult males develop enormous cheek pads that stick out from the side of the face like blinkers on a racehorse; below the chin is an ample, loosely-hanging flap of skin which contains an aïr-sac having a similar function to that described in the siamang. The call of the orang, seldom if ever heard in captivity where they are regarded as very silent animals, is a bellow both loud and long ending in a high-pitched squeal. Sumatran and Bornean orangs have quite different facial characteristics. The Bornean males have a fatter face, more prominent jaws and cheek pads which project forward, like blinkers; the face of the Sumatran orang is long, oval, and flatter with a prominent mustache and beard; the cheek pads project sideways, in the same plane as the face. One authority has described the Bornean form as looking like an Indian Buddha and the Sumatran as resembling a Chinese mandarin.

Orangs are thought to be wholly arboreal, a surprising conclusion in view of their large bulk, living largely on fruit and leaves. Their bodily structure is similar to that of all apes, inasmuch as the arms and hands are enormously elongated and the legs are relatively short. They differ from chimps and gorillas, whose limbs have undergone secondary ground-living adaptations, by evolving a remarkable degree of mobility of the hip which has become almost as flexible as the shoulder. The legs can be placed at the most bizarre angles as the animal climbs amongst the branches of the trees. That this behavioral character is genetically induced is substantiated by the anatomy of the hip joint; orangs differ in a number of morphological characters from the African apes, i.e., the absence of the internal ligament of the hip joint and the peculiar (almost man-like) orientation of the lesser trochanter of the femur.

Little is known about the social life of orangs in the wild as they are few in number and difficult to observe, but they are generally assumed to live in small family groups.

Finally we come to the family of man (Hominidae). Today this family consists of a single polytypic species of world-wide distribution. Fossil evidence indicates that in the past there have been many different species of man, all of which are now extinct. The key to the formation of species is geographic isolation, and no portion of mankind enjoys (or has enjoyed for a very long time) the degree of true isolation necessary for speciation to take place. Human populations may become isolated temporarily for geographical, political, or ideological reasons, but sooner or later populations overlap at their edges and the genes once more start to mix.

Man's physical characters and evolution are the principal subjects of the last three chapters of this book and so need not be dwelt on here.

CHAPTER FOUR

Habit and habitat

Habit is what an animal does and habitat is where it lives; and what an animal does depends very largely on where it lives.

There are three levels at which an organism can be studied: the structural, the functional, and the behavioral. A simple analogy may help to define the elements of this triple-decker concept. Suppose, as a Do-it-Yourself addict, you see an advertisement for a Kar-Kit to build your own automobile. When the kit arrives you unpack it and lay out the components on the garage floor. With the help of the instruction book you identify each separate item, thereby acquiring a knowledge of the anatomy of the car or, in other words, its *structure*. Once you are familiar with the various parts, you can start assembling them system by system until finally the job is done and there, in all its wide-tracking glory, is the final product. Now is the time to fill up with gas and oil and switch on the ignition. The engine fires and, with a little throttle, idles nicely—the car is showing perfect *function*. But you are not yet in the clear for the final test—the only test in fact—of your investment is *behavior*. Your automobile may be composed of high quality parts, which work together immaculately in your garage, but how does the car match up to the real world,

Table 2 **Primate Distribution in Terms of Vegetational Zones**

	Africa and Madagascar	India and S.E. Asia	Central and South America
Tropical Rain Forest	Arctocebus Avahi Cercocebus Cercopithecus Cheirogaleus Colobus Daubentonia Galago Gorilla Hapalemur Indri Lepilemur Lemur Mandrillus Microcebus Pan Perodicticus Phaner Propithecus	Anathana Cynopithecus Dendrogale Hylobates Loris Macaca Nasalis Nycticebus Pongo Presbytis Ptilocercus Pygathrix Simias Symphalangus Tarsius Tupaia Urogale	Alouatta Aotus Ateles Brachyteles Cacajao Callicebus Callimico Callithrix Cebuella Cebus Chiropotes Lagothrix Leontideus Pithecia Saguinus Saimiri
Tropical Montane Forest	Cercopithecus Colobus Gorilla Perodicticus	Cynopithecus Dendrogale Hylobates Macaca Presbytis Rhinopithecus Symphalangus Tupaia Urogale	Alouatta Aotus Ateles Brachyteles Cebus Lagothrix Leontideus Saguinus
Monsoon Forest		Anathana Hylobates Loris Macaca Nycticebus Ptilocercus Presbytis Symphalangus	

to the environment—its environment, the highway? So, nothing loath, you take the next step and introduce it to its habitat. Then, and only then, are you in a position to judge its effectiveness. The success of your investment depends on the level of its roadworthiness, its behavior in terms of turning, gas consumption, and so on. Thus habit (or behavior) and habitat (or environment) and habitus (or structure) are very properly studied together whether one is concerned with macaques or Cadillacs.

Clearly if the nature of the environment plays such an important role in evolution, we must do more than pay lip-service to it—we must study it. Instead of asking ourselves why certain primates living in a particular habitat have a certain structure or behavioral quirk, it is frequently rewarding to turn the whole problem upside down and ask what

Table 2   **Primate Distribution in Terms of Vegetational Zones—Continued**

	Africa and Madagascar	India and S.E. Asia	Central and South America
Mixed Deciduous Forest	Arctocebus Cercopithecus Cheirogaleus Galago Lemur Microcebus Pan Papio Perodicticus Propithecus	Anathana Hylobates Loris Macaca Presbytis	Alouatta Aotus Cebus
Thorn Forest	Cercopithecus Lemur Papio Propithecus	Anathana Macaca	
Savanna Woodland	Cercopithecus Erythrocebus Galaga Lemur *(L. catta)* Pan Papio	Loris Macaca Presbytis	Callithrix
Open Savanna	Cercopithecus (*C. aethiops*) Erythrocebus Papio	Macaca Presbytis (*P. entellus*)	
Steppe	Erythrocebus Papio Theropithecus		

adaptations a given type of animal in a given habitat is likely to need. To answer this, one must turn one's attention from the animal to its ecology—to the nature and properties of the environment. By such methods adaptations become apparent that have not been recognized before and the functional significance of others, previously regarded as "nonadaptive," become strikingly obvious. At this point, definitions of "environment" and "ecology" might be useful. The environment is the aggregation of all physical, chemical or biological factors that impinge upon an animal and ecology is the science that deals with the relationships between an animal and its environment. The totality of animals and plants in a given environment is called an "ecosystem" and the "ecological niche" is the particular zone for which the animal is peculiarly adapted.

In spite of the dramatic success of certain ground-living primates as *Homo, Papio,* and *Macaca,* the primate way of life is essentially an arboreal one. Of the 189 living species of primates 166 are wholly or partly arboreal (see Table 2).

Man and the baboons are the break-away exceptions that prove the rule. Morphologically, however, these species have for the most part retained their primitive arboreal characters. The anatomical-physiological characters of ground-living primates are still characteristic of the adaptations for the three-dimensional arboreal life of their remote ancestors. In the face of the largely two-dimensional demands of life on the ground, some tree-climbing adaptations have become redundant and others have been deployed afresh in new directions. Few specifically terrestrial adaptations have made their appearance. It is only when one turns to behavioral adaptations that the major effects of life on the ground become apparent.

In recent years the advances in primate ecology and ethology have demonstrated how closely behavioral differences are linked with habitat, to the extent that behavioral variability is not only different between genera but also between different species of a single genus (Table 3). The Hamadryas baboon (*Papio hamadryas*) of the highlands of Ethiopia and of Somalia differ in their social organization from the olive baboons (*Papio*

Table 3 **Behavioral and Ecological Aspects of the Behavior and Environment of Ground-living African Monkeys**

	Patas (*Erythrocebus*)	Gelada (*Theropithecus*)	Hamadryas Baboon (*Papio hamadryas*)	Common Baboon (*Papio*)
Habitat	Open savanna, trees & grass	High montane, moorland	Arid savanna, few trees, grass	Woodland & open savanna; trees & grass
Reproductive unit	One-male group (Harem)	One-male group (Harem)	One-male group (Harem)	Multi-male group
Male dominance	No male dominance	Male dominance without "neck-bite"	Male dominance with "neck-bite"	Male dominance without "neck-bite"
Predator response	Non-aggressive Hide in long grass	Aggressive; threat behavior	Aggressive; threat behavior	Aggressive; threat behavior
Group size	Mean: 15 individuals	Herds of up to 400 individuals	30-50 individuals	40-80 individuals
Births	Seasonal	Seasonal	Seasonal	Seasonal
Sexual dimorphism	Marked	Marked	Marked	Marked
Sleeping sites	Trees	Steep cliffs	Steep cliffs	Trees or cliffs

anubis) of the lusher East African grasslands which, in turn, are found to behave differently from the forest-living races of Uganda (*Papio anubis*) and from the chacma baboons of Cape Peninsula (*P. ursinus*). There are almost as many behavioral patterns amongst baboons as there are vegetational and climatic zones.

The somewhat naive assumption among primate biologists that non-forested tropical environments can be lumped into simple grassland or savanna biomes has tended to oversimplify our thinking about primate ecology both in contemporary and historical terms. The growing awareness of the variety of ground-living habitats available to primates has produced a much deeper understanding of the variation in social behavior that can be seen in zoologically linked, but ecologically discrete, groups. There is some evidence, too, of seasonal variability in social behavior. In the case of the gelada baboon this is related to periods of food scarcity when the "herds" comprising as many as 400 animals break up into one-male units consisting of an adult male and several females. Clearly the one-male grouping system is of adaptive value as it guarantees that females can get a fair share of the available food (Crook, 1966).

Seasonality and its effect on behavior has been studied in Japanese macaques, some groups of which live in the Shiga Mountains on the island of Honshu where winter snows last for four or five months. Changes are seen in dietary habit, in grooming, in locomotion, and in normal spacing between individuals. The latter behavior called "clumping" occurs as a number of individuals closely embrace each other to protect themselves against the cold. New subcultural habits develop and spread rapidly within a troop subjected to extreme hardship. Suzuki (1955) describes how one troop adopted the practice of bathing in hot springs one winter, gave it up in the summer and readopted the practice the following winter.

Students of primate behavior in the wild have principally concerned themselves with ground-living forms for the good reason that troops are more easily followed and observed in open country than in high forest. Consequently we have little knowledge as yet of the significant micro-habitats of forest-living primates or of the extent to which social systems vary between arboreal species. The howler monkeys are the best known of the arboreal monkeys, thanks to the extensive studies of *Alouatta villosa* by Carpenter and others who have followed him on Barro Colorado Island, Panama. Field studies of gibbons (Carpenter, 1940; Ellefson, 1968), chimpanzees (Van Lawick-Goodall, 1969, and Reynolds and

Reynolds, 1965); colobus monkeys (Schenkel and Schenkel-Hülliger, 1967), langurs (Jay, 1965, and Ripley, 1967), mona monkeys (Bourlière, Bertrand, and Hunkeler, 1969), drills (Gartlan, n.d.), talapoins (Gautier-Hion, 1966), and mangabeys (Chalmers, 1968) have already provided a foretaste of the diversity apparent in a single major habitat (forests) comprising a seemingly infinite number of distinctive ecological niches.

MAN'S DUAL HERITAGE

When the behavior of arboreal primates is as well known as that of ground-dwellers, it is quite certain that the ethologists, social scientists, and psychologists will be amazed to discover what some anthropologists have been aware of all along—that the root of man's genetic inheritance is bifid, derived in equal parts from his phylogenetically older, arboreal "infancy" and his recent ground-living "adolescence." Man comprises a fascinating mixture of the timid and the aggressive, of the "flight" responses of forest animals and the "fight" responses of the savanna forms; the family allegiances of the gibbon and the marmoset and the community consciousness of the baboon; the arms and trunk of an arboreal ape and the legs and hands of a ground-living monkey; and a mind that is uniquely his own but derived, nevertheless, from a composite of psychological and neurological adaptations to the demands of both forest and savanna life. Man reveals his distant arboreal ancestry in times of chronic stress. Drop-outs from society for whatever cause "take to the woods" because here is the security, the anonymity, the natural food supply that promises survival. The little cabin in the woods is every man's romantic idea of escapism. Man's mythology contains countless references to woods and forests; we both love and fear them—as exemplified by the immortal drawings of Arthur Rackham which perfectly express this duality. The tales of Snow White, Robin Hood, and of the Swiss Family Robinson, of Winnie-the-Pooh and of the legendary wild man of modern times—the Yeti, the Bigfoot, and the seemingly endless stream of hairy woodsmen—living so it seems to urbanized man in an enviable and bucolic dream world. Man, stripped down by circumstances to a psychological base-line, returns to his remoter phylogenetic past where security from predators, multiple escape routes, an abundant supply of natural food, and a solitary existence meet his needs. The savanna is for the fighter, the woods for the recluse. We can recognize both these conflicting pressures in our own way of life. We work in the city but commute at night to our homes in the suburbs, the nearest most of us can get to the forest. We relish the contrast but we don't always recognize the biological roots of our satisfaction.

PRIMATE ECOLOGY TODAY

With very few exceptions, living non-human primates are restricted to tropical latitudes occupying three major habitats, forest, woodland-savanna, and grassland, between 25°N and 30°S (Figure 5). Within this belt they have a considerable altitudinal range, an aspect of their environment not usually recognized. A few examples of high mountaineers are listed in Table 4.

A few primate species occur naturally in temperate latitudes. For example *Macaca mulatta* is found near Peking, and *Macaca fuscata* on

Table 4   **High Mountaineers Among Primates**

Genus	Region	Maximum Altitude (feet)*	Latitude
Theropithecus	Ethiopia	16,400	14°N
Gorilla	Virunga Volcanoes	13,500	4°S
Macaca	Szechwan, China	13,000	30°N
Presbytis	Himalayas	12,000	28°N
Tupaia	Yunnan, S.W. China	11,000	27°N
Rhinopithecus	Szechwan, China	10,000	30°N
Pan	E. Africa	10,000	0°
Aotus	N. Colombia	9,000	5°N
Cercopithecus	Uganda, E. Africa	10,000	0°
Alouatta	Venezuela	8,200	10°N

*Altitudes given represent the upper limit of the range.

Honshu Island, Japan, at 41 °N. During the Pleistocene interglacials, the range of *Macaca* extended as far north as the British Isles. The principal factor limiting the northward spread of ground-living primates is likely to have been dietary rather than climatic. The seasonality of fruit and leaves, the difficulty of digging grasses, roots, and tubers from the frozen ground during the winter imposes an ecological threshold on the northerly range of monkeys. Macaques are hardy animals which withstand cold climates in captivity extremely well as long as food is being supplied. A colony of *M. mulatta* has been maintained in outdoor conditions in a suburb of Moscow since 1963 surviving even subzero winter temperatures. Savanna monkeys (*Cercopithecus aethiops*), too, have been kept successfully under the same conditions. However, the Japanese macaques of the Shimokita Peninsula, the most northerly part of the Japanese island of Honshu, survive the snowy winters where temperatures may reach as low as $-4°$ F, subsisting for several months almost wholly on bark.

It would appear that climate alone would not have been a sufficient deterrent to the northerly migration of Old World monkeys.

Zoogeography   The main areas of primate distribution (Figure 5) are Africa (including Madagascar), India and Southeast Asia, and South and Central America.

Africa is the home of the prosimians, Old World monkeys of both cercopithecoid subfamilies—Colobinae and Cercopithecinae—and of apes, the gorilla and the chimpanzee. Madagascar is the only home of the lemurs. How these animals reached Madagascar and how they thrived, and later, suffered the coming of man has been told by Alan Walker (1967). It is a dramatic story, but at the same time a sad one for these beautiful and rare creatures are one of the first major victims of man's capacity for destructiveness. Conservation problems, so pressing today, started in Madagascar over 3000 years ago when man first colonized the island and destroyed all the giant, slow-moving, ground-living forms, leaving only the small and agile arboreal lemurs to survive to tell the tale. Thus, today, we see not a natural assemblage of lemurs in Madagascar but an unbalanced fauna in which nocturnal genera predominate over diurnal ones and ground-living forms are absent. The ground-living niche has been partly filled by some of the true lemurs particularly *Lemur catta* (the ring-tail lemur). The ring-tail has been called the "baboon" of Madagascar on account of its allegedly ground-living habits but this is somewhat of a misnomer according to Alison Jolly (1967).

The non-Madagascan prosimians found in Africa are the pottos and galagos, both nocturnal forms that have survived the competition of the African monkeys by taking ecological refuge in the fastness of the night.

The only primate to be found in Europe, the Barbary ape, doesn't really belong there at all. The Barbary apes, inhabitants of the Atlas mountains of North Africa, were already inhabitants of Gibraltar in 1704 when the British captured the island during the war of the Spanish Succession. When Gibraltar was ceded to Britain by Spain in 1713 under the Treaty of Utrecht the monkeys were established as permanent residents and added to the pay-roll of Her Majesty's Forces. The traditional British fondness for animals must, presumably, be credited with this zoogeographically confusing procedure. Barbary apes are macaques and the *only* macaques with a foothold in Africa. Their presence there strongly suggests a Pleistocene ice-age invasion, for modern macaques are essentially Asian forms; their ecological counterparts in Africa are the baboons which I believe are derived from a macaque-like stock that entered Africa via the Afro-Asian land bridge during the late Miocene epoch (see p. 86).

The primate stocks of India and Southeast Asia comprise the same families as those of Africa—Cercopithecidae, Pongidae, Lorisidae. In addition this realm possesses three families unknown in Africa, Hylobatidae (the gibbons and siamangs), Tarsiidae (the tarsiers), and Tupaiidae (the treeshrews). There are puzzles here, too. The island of Celebes lies across the Makassar Straits from Borneo; the Straits are the site of a deep marine trench that could never conceivably have been connected by land to Borneo or Sumatra during the Cenozoic. There are three primate genera on Celebes: *Tarsius, Macaca,* and *Cynopithecus.* How did they get there? From Borneo by "rafting"? Or from the Philippines via the Sangi Island chain? The macaque is undeniably a macaque, but *Cynopithecus*—the so-called Celebes "ape"—is a what? In many ways it is macaque-like but in others it is baboon-like; this is a bit worrying as the nearest "baboon" is 5,600 miles away in Africa. The Celebes baboon may well be an example of environmental selection producing parallelism in geographically remote areas. Baboons, it has been said, are not so much a genus—they are more a way of life. Given a similar environmental situation, similar forms arising from common evolutionary stock will become physically adapted in a similar fashion. There are many zoogeographical puzzles in Southeast Asia and there are many species of "unknown" primates. China is the home of *Rhinopithecus,* the Chinese mountain langur that is quite unknown in the West as a living creature, never having been exhibited in a zoo outside China. *Pygathrix,* one of the langurs of Vietnam, is almost equally mysterious. Then there is *Simias,* also a langur, which lives in splendid seclusion on the tiny, marshy Mentawi Islands off the west coast of Sumatra. It must be one of the few animals in the world of which no photograph is known to exist.

Although the North American continent is the home of some of the earliest fossil primates known, there are no primates (non-human) north of latitude 24° N. The southernmost limit of their present distribution is 30° S. Between these tropical latitudes there are an incredible variety of arboreal monkeys, but there is not one single prosimian, not one single ground-living monkey and not one single ape. Why? The answer undoubtedly lies in the ecological differences between the Old and New Worlds but we are nowhere near understanding at present what these differences are.

Two geographical features which have dominated the pattern of primate distribution in South America are the coastal range of the Andes mountains, a significant barrier to westward primate migration, and the Amazon river. The pattern of the Amazonian drainage system has divided the heartland of South America into a series of enclaves, north and south of the river, each of which is bounded on one side by the east-west course of the Amazon and its continuation the Solimoes, and in a north-south direction by the Amazon's multiple tributaries. This geographic feature has resulted in a segmentation of South American primates, notably the marmosets and tamarins, into discrete populations that show a bewildering array of variations in coat color, facial patterns, adornments, ear-tufts, and mustaches. This is just one example of the profound effect that South American geography has had on primate distribution and speciation.

Major Habitats In order to rationalize the morphological adaptations and behavioral components of the primate way of life, it is essential to view primates in their natural biotic provinces, the floral and faunal communities in which they live and form a part. A note of caution may be in order here. With certain reservations one assumes that present habitats of primates are comparable to their past habitats. And, therefore, one further assumes that it is meaningful to look for the selective advantage of certain morphological and behavioral patterns, in the context of present environments. It must be borne in mind, however, that present-day vegetation is not always representative of the natural climax of the region. The effects of fire, agriculture, cattle-grazing, logging and road-making are apt to create novel, vegetational types that were unknown before the coming of man. Although botanists disagree on what is man-made in vegetation and what is not, they would probably be in accord with the general statements that man has transformed much of the rain forest of Africa and Madagascar into grassland, the grassland into steppe, and steppe into actual desert.

Man's chief weapon has been fire. In Tanzania, Rhodesia, and Zambia, man-made fire set for agricultural or for grazing purposes is responsible for the degradation of thousands of acres of woodland. Slash-and-burn techniques associated with shifting agricultural patterns have devastated vast areas of Malaya, the East Indies, and the *llanos* of South America. It has been said that fire is a more powerful instrument in the hands of primitive peoples than the hoe. The second major force for degradation has been overgrazing by domestic cattle particularly in Africa and India. The depredations of goats in North Africa are well known to have reduced areas of dry deciduous forest to desert. Indiscriminate logging can so denude mountainsides that with the next heavy rain the top soil peels off and slides into the valley; yet another aspect of man's impact on natural vegetation in his hunger for wood.

One would be in serious error to assume that secondary forests and grasslands did not exist before the advent of man. Natural fires set by lightning must always have been a hazard. The open, treeless prairies of North America are thought to have resulted from the combination of seasonal drought, dry grasses, high winds—and fire (Wells, 1965). The savannas of tropical South America, which are virtually devoid of grazing animals (wild or domestic), show little evidence of human-induced fires. Beard (1953) expresses the opinion that the savannas of South America are, and always have been, the climax vegetation. Yet, a question remains. Why, if the South American grasslands are so ancient and natural, have so few species of mammals become adapted to a grassland environment? There are no South American savanna primates, ruminants, or carnivores—the contrast to the fauna of the African grasslands is striking.

In a classification of contemporary primate habitats, it is really irrelevant whether or not communities are artificial but, of course, it is of great importance to distinguish them when primate ecology is being considered in an evolutionary context.

Vegetational Zones   Primates are found in three principal vegetational communities: (1) tropical forest, (2) woodland savanna, and (3) open savanna (Table 2).

Tropical forests vary in their physiognomy according to the factors of altitude, latitude, and rainfall. Rainfall throughout the year, at zero altitude and at zero + or −5 degrees latitude produces the true tropical rain forest, which is evergreen and nonseasonal (Figure 7). Trees in areas with less rainfall shed their leaves to conserve water; in tropical forests there is plenty of water and the forests remain evergreen. Leaves are shed continuously and are rapidly replaced so that there is never a "season" when the trees are leafless. Fruiting trees are to be found in one part or another of the forest nearly all the year round.

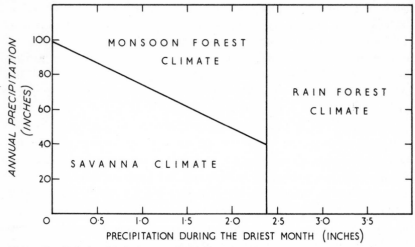

Figure 7.   Relationship between rainfall and vegetation.

The chief physiognomic features of rain forests are evergreen broad-leaved trees forming a closed canopy, which may be so dense as to cut out all sunlight from reaching the forest floor; herbaceous epiphytes and thick-stemmed lianes drape the trees linking one crown with the next so as to form a dense horizontal and vertical network. Trees with buttress roots that have both the appearance and the function of architectural buttresses are characteristic.

Three strata of the canopy are usually recognized by botanists—an understorey, a middle storey, and an upper storey. The understorey is very often closed, that is to say the crowns overlap one another and form a dense—virtually continuous—horizontal layer. The middle storey is characterized by crowns of trees that are sometimes in lateral contact but do not usually form a closed layer. Above this is the upper storey where the crowns of giant trees soar to 150 feet above the forest floor. The crowns of this stratum are seldom in contact. In addition to the three strata an occasional forest giant rises above the upper storey. Not all rain forests are

76

stratified and the stratification itself is usually very difficult to observe unless the forest has been cut through by a road or a river. Figure 8 illustrates stratification in a very stylized way; three storeys are apparent but, to simplify the description of the vertical distribution of primates within the canopy, only two strata are identified. The under-storey and middle storey

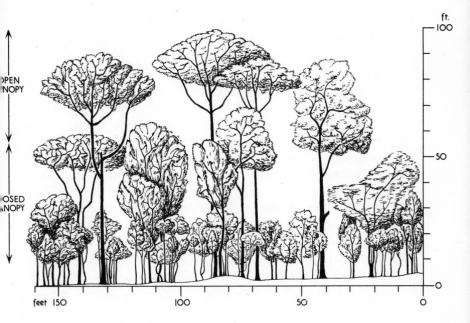

Figure 8. Stratification of tropical rain forest in West Africa. (Adapted from P. W. Richards, *The Tropical Rain Forest*, Cambridge University Press, 1966.)

can together be referred to as "closed" canopy and the upper storey as "open" canopy.

Other types of tropical forest inhabited by primates are Monsoon Forest, Dry Deciduous Forest, Thorn Forest, and Tropical Montane Forest. The first three types form a series in which the annual rainfall is progressively diminished: 40-80 inches for Monsoon Forest with two or three dry months in the year, 20-50 inches for Dry Deciduous Forest with a prolonged dry season, and less than 30 inches a year for Thorn Forest. Theoretically monsoon areas with a long winter drought should not be forested. However, sufficient moisture accumulates in the soil of the rainy season to maintain the trees throughout the year. Most monsoon forest

77

trees are deciduous and conserve moisture by shedding their leaves in the dry season.

A tropical montane forest is a rain forest at high altitudes (between 2000-3000 feet above sea level). The typical three storeys of tropical rain forest are reduced to two, the trees are lower and tree-trunks shorter. With increasing altitude, forest gives way to thickets which may consist entirely of stands of bamboo, and then to the quaintly named elfin woodland, and finally to montane heathland which in Africa is characterized by giant groundsels and lobelias. Rainfall is high and the atmosphere is intensely humid and misty, conditions which have prompted botanists to give montane forest the descriptive name of "cloud forest."

Tropical grasslands are a mixture of grasses and trees. The progression from closed woodland to open grassland is characterized by a decrease in the height of the trees and an increase in the density and the height of the grass. Woodland savanna has a seasonal rainfall of about 30-50 inches per year, while in the open savannas (or true tropical grasslands) the rainfall drops to less than 30 inches. When it is less than 20 inches per year the grasses become short and tussocky and the trees widely scattered and stunted. Gradually steppe grades into desert. Once again, the critical factor is rainfall (Figure 7).

Woodland savanna contains low deciduous trees with their crowns in lateral contact; the canopy is therefore single-storeyed and semi-closed allowing plenty of sunshine to strike the forest floor where grasses, shrubs, and herbs can grow. As woodland grades into savanna the cover becomes lighter, the trees becoming spaced further apart. Eventually open savanna supervenes, the grasses become the dominant form of vegetation. In parts of Africa with high rainfall, the grasses may grow from 5-15 feet in height. The characteristic seasonal events of African grasslands are the annual fires which burn the grasses and saplings, leaving the fire-resistant acacias as the only surviving vegetation. With the coming of the rains the new grasses grow rapidly.

PRIMATE ECOLOGY IN THE PAST

The past of organisms is one of the determinants of their future, so primate characteristics should be viewed in the framework of evolution as well as in the sharp light of the present because it is in the evolutionary context that relationship between structure and behavior becomes most meaningful.

The morphological characteristics of primates are the outcome of certain evolutionary trends discussed in detail by Clark (1959) and by Napier and Napier (1967). These trends (see p. 40) did not develop simultaneously nor did they proceed at the same evolutionary rates. At any given moment in time it is possible—through the study of paleontology—to see primate trends at various stages of development. Observed as it were in isolation, the adaptive significance of these trends is more easily apparent if the fossil primates themselves are set in the framework of ecology. It is, then, possible to determine certain broad ecological principles that have governed the adaptations of the Order, their dispersal, and their taxonomic differentiation.

Just as studies of living animals are meaningless in the absence of environmental information, so evolutionary studies are of doubtful value without a background of historical ecology—of paleoclimatology, paleobotany, and paleogeography. There are many puzzling characteristics of primate morphology and behavior that seem inexplicable in the light of the present, but become comprehensible when viewed in historical perspective.

There are, for example, a number of puzzling locomotor habits among living primates—the slow-climbing of lorises, the vertical clinging of tarsiers and of certain Madagascan lemurs, and the knuckle-walking of chimpanzees and gorillas. There are also a number of physical characteristics that do not quite fit with the present-day environment of the animal concerned: the "sexual skin" of talapoin monkeys, the cheek pouches of guenons, the thumbless hand of the colobus monkey, the ground-adapted limb proportions of the arboreal proboscis monkey, and the peculiar conformation of the incisors and canines of the South American subfamily Pitheciinae. The significance of these adaptations is not easy to understand in the light of present-day ecology; perhaps this is simply because of ignorance. But it may also be that these structural and behavioral patterns have persisted long after the selection pressures that originally brought them about have ceased to operate. They have become atavisms, like the appendix or the lobe of the ear in man. It may be these characters have become irrelevant, but neutral, and are retained in the absence of any selective force working against them. It also may be that they have assumed a *secondary* function and have a present use quite unrelated to their past use. Many behavioral characters used in social and sexual contexts by animals are believed to have evolved in this way. A possible case of behavioral atavism in primates is Kortland's "dehumanization" theory. Adriaan Kortlandt, a Dutch zoologist, regards the use of sticks by wild chimpanzees against predators (leopards, for instance), their use of tools, their obvious intelligence, their frequent recourse to two-footed walking, and their occasional carnivorous habits, as indications that chimpanzees were one-time dwellers in open savannas, an environment in which they are seldom seen today. The possession of these abilities, usually considered to be hallmarks of man, indicates to Kortlandt that chimpanzees were once more "human" than they are now. With the coming of man and his success as a hunter, chimpanzees were forced *back* into the forests where they are found today; they have become "dehumanized," but retain some of their past patterns of behavior such as the use of weapons against predators. It would seem to be a case of special pleading to suggest that chimpanzees have retained certain patterns of behavior as relics from the past when a far simpler explanation is available—that these patterns have been acquired in response to present needs.

It must be remembered too that today we are looking at an exceedingly narrow slice of primate time. When the last ten years, during which so much of our present knowledge has accumulated, is expressed as a proportion of the length of the Cenozoic (the period during which primate evolution took place), it becomes apparent that it represents 0.000015 percent of the total time since the primates set out on their evolutionary course. It is quite probable that the present habitat of a given species is not the same as it was when a particular adaptation evolved. Much of the non-forested areas of Africa, for instance, owe their present form to the effects of fire, agriculture, cattle grazing, and logging. The "natural" habitat of the rhesus macaque in many parts of central India where it is regarded as sacred are the temples, the rooftops, and the abandoned buildings of towns and villages. There is a considerable danger in attempting to establish too close a correlation between habit and habitat unless there is the assurance, firstly, that the vegetation represents a community that is in dynamic equilibrium with the prevailing climate (climax vegetation) and, secondly, that the particular primate population is strictly indigenous.

It is a measure of the generalized structure and the physiological adaptability of primates that various populations of these animals have suffered relocation in alien environments without apparently prejudicing their survival. To name a few expatriate primates, there are the green monkeys (*Cercopithecus sabaeus*) of St. Kitt's in the West Indies, the mona monkeys (*C. mona*) on Grenada, also in the West Indies, the crab-eating macaques (*M. fascicularis*) on Mauritius, the rhesus macaques on Cayo Santiago Island off Puerto Rico, and *M. sylvanus,* the Barbary macaques on Gilbraltar. Finally, there is even said to be a colony of squirrel monkeys living wild in Florida, U.S.A. Many of the populations of primates that we now regard as native because they seem to be approximately in the right place, may equally well be newcomers for all we know; only through zoogeographical information obtained from paleontology will these sorts of questions be answered.

In order to understand the zoogeographical distribution of primates and the nature of the life zones in which they are now living (Table 2) it is necessary to review briefly the geographical and climatic changes that have occurred during the 65 million years of the Cenozoic, the era during which primates evolved. The Cenozoic era is divided into epochs; their names and durations in millions of years are shown in Figure 9. The Cenozoic, a zoological term meaning recent forms of life, denotes the period from the end of the Cretaceous to the present day. Geologists use the term Tertiary to cover much the same period; however the Tertiary era finishes at the end of the Pliocene when the Quaternary era takes over. Most people find that it is almost impossible to conceive of a million years, let alone 65 million years of primate evolution or 600 million years that have elapsed since the chordate forerunners of the vertebrates first appeared on earth. Think of the time that has elapsed since the origin of the chordates as a year of our own time, a calendar year divided into months, days, weeks, hours, and minutes. The first true vertebrates appeared on March 23rd, the first mammals on September 25th. The primate order evolved on November 27th, the first apes came into being on December 12th, but it was not until 8:15 p.m. on December 31st that modern man (*Homo sapiens*) stepped across the threshold. This is a sobering thought for us in our moments of anthropocentric exuberance.

Since the beginning of the Cenozoic, world temperatures have been steadily dropping (Figure 10). During the Eocene the mean annual temperatures in Seattle, Paris, and London were comparable to those experienced today in Mexico City. Subtropical forests extended as far north as 53° latitude. At the end of the Eocene, the New Siberian Islands well within the Arctic Circle, enjoyed a temperate climate as evidenced by the nature of the fossil plants (Figure 11). During the Oligocene there was a sharp acceleration in the cooling process; subtropical forests reached no further north than Los Angeles. In western Europe warm temperate forests had replaced tropical forests and yearly mean tempera-

GEOLOGICAL TIME SCALE	
10	PALEOCENE
18	EOCENE
11	OLIGOCENE
14	MIOCENE
10	PLIOCENE
2-3	PLEISTOCENE

Total duration: approx. 65 million years

Figure 9. Geological epochs of the Cenozoic. Numbers indicate millions of years duration.

tures were in the neighborhood of 64°F (18°C). In the Miocene the climate of California was warmer than today, but not subtropical. By the upper Miocene average temperatures of the world were much as they are at present but the climate was more equable, there being less contrast between summer and winter. By the Pliocene, permanent polar icecaps had formed and arctic weather was moving southwards, bringing conditions for glacier formation—heavy winter snowfalls and cool, cloudy summers—into the middle latitudes.

During the latter half of the Pleistocene the polar ice extended far southwards and brought arctic conditions into regions of the Northern Hemisphere that are now temperate. The ice-sheets, unable to melt in the cooler summers, crept down from the north pre-empting unaccountable tons of water which appreciably lowered sea levels throughout the world and raised previously submerged continental shelves linking Iceland to Britain, Britain to Europe, and Europe to Africa. Tundra conditions, associated with mammals of cold regions such as the musk ox and walrus, extended into California, Texas, and Mississippi.

Between the four great Ice Ages—the Gunz, the Mindel, the Riss, and the Würm[9]—were long intervals or "interglacials" when warmer conditions prevailed and when mammals, including non-human primates and man, swarmed back into the then temperate regions of Europe, Asia, and North America from whence they had been driven by the encroaching ice-sheets.

The effect of these climatic changes on the distribution of primates was considerable. In the Eocene subtropical and tropical forests spread 50° North and South of the equator providing a wide habitable belt of 100° latitude; today the belt of subtropical and tropical vegetation—the home of primates—has shrunk to less than 50° latitude (Figure 11). Many of the areas between the tropics of Cancer and Capricorn are either deserts, grasslands, mountain regions, or high plateaus; there is very little actual forest. The primate world has shrunk since the days when primates abounded in the vast forests of North America, Europe, Asia, and southern England. The actual area of forest habitat presently available is probably less than 5 percent of what it used to be 60 million years ago.

It is of particular importance for the study of primate taxonomy and

---

[9] The glacial periods in Europe and America were approximately contemporaneous, but in North America they are named the Nebraskan, the Kansan, the Illinoian, and the Wisconsin, respectively.

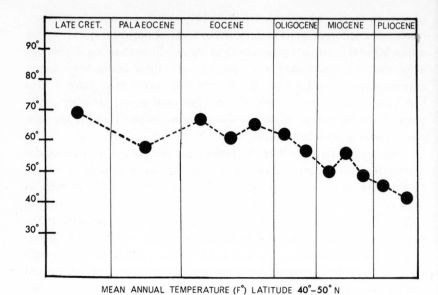

Figure 10.   Mean annual temperature changes during the Tertiary. (Adapted from Earling Dorf, "The Earth's Changing Climates." *Weatherwise,* 10 (1957): 54.)

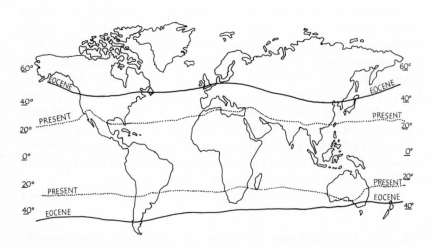

Figure 11.   Limits of tropical and subtropical climates in the Eocene (solid line) and the present (dotted line).

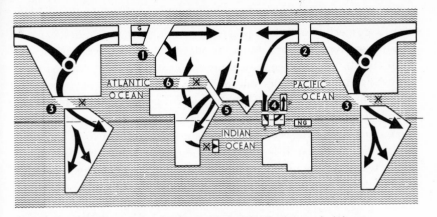

Figure 12. Migration routes of primates during the Tertiary period. Large black dots represent centers of dispersal and arrows, the routes. Numbers indicate sites of land bridges as follows: (1) North Atlantic land bridge, (2) Bering Strait land bridge, (3) land bridge between North and South America, (4) the Sunda Shelf land bridges, (5) Afro-Asian land bridge, (6) probable site of Afro-European land bridge.
Broken line represents the site of the Uralian Trough, the northern arm of the Tethys Sea. ✕ indicates sites of possible raftings.

evolution to know something about the migrations of primates during the Cenozoic and to determine when and by what routes they reached the areas in which they are now to be found. In Figure 12 the solid black spots represent centers of dispersal, the arrows—the routes; the numbered spots indicate land bridges or sea routes which have been open at various times in the last 65 million years. It can be seen that, in the absence of other evidence, North America is provisionally regarded as the birthplace of primates. From North America they reached Europe and Asia by two routes: *via* the North Atlantic bridge to Europe and *via* the Bering Straits to Asia. Primate migrations into Europe and Asia could only have occurred in the early part of the Tertiary when the Asian and European land bridges were still warm and forested enough to permit the passage of primates used to tropical conditions. It is unlikely that there were any further movements of primate stocks between Asia, Europe, and North America after the middle Eocene. From North America, primates reached Central America via a land connection, but the passage to South America was blocked by a broad seaway which separated the two continents from the middle of the Eocene until the Pliocene. Nevertheless, it is certain that a crossing did take place as fossil primates are found in South America dating from the Miocene. The likely mechanism is by a process of "rafting" which implies a chance crossing of rivers or open water by means of floating vegetation, tree trunks or blocks of matted grass roots. Animals have been observed upon such "islands" many miles out to sea. G. G. Simpson has given this method of migration the felicitous name of the "sweepstake route," epitomizing the essential ingredient of chance (Figure 12). Another sweepstake route crosses the Mozambique channel from Africa to Madagascar; this route is believed to have carried the earliest lemurs to Madagascar, where they have been evolving independently since the Oligocene. Other land bridges shown in Figure 12 include the Afro-Asian land bridge which connected Africa to northern India and the Far East during most of the Miocene epoch and which allowed an exchange of early Old World monkeys and apes between the two continents. There is reason to believe that early hominids such as *Ramapithecus* reached Asia from Africa via this route (see Chapter 9). A land connection between Africa and Europe in the region of Sicily or further west across the Straits of Gibraltar probably existed for short periods during the Oligocene and Miocene. Prior to these epochs Europe and Africa were separated by the broad Tethys Sea of which the Mediterranean is now a tiny remnant. So much mountain building was going on in the bed of the Tethys that a brief connection could well have been established in the late Eocene via a string of islands. Migration of primates from Europe to Africa may, therefore, have occurred by a series of sweepstake routes, the animals hopping from island to island along the chain.

Land bridge connections between Southeast Asia and the islands of Sumatra, Java, and Borneo were frequent occurrences as the sea rose and fell, especially during the Pleistocene; all these islands lie within the 100-fathom mark of the Sunda Shelf. Human and non-human primates must have used these bridges to colonize island after island ultimately reaching the Philippines along two narrow island chains stretching from northeast Borneo to Mindanao and Luzon.

Apart from the climatic and vegetational changes, there were other revolutions going on during the Cenozoic. The face of the land surface was also changing. The vast mountain ranges of the world today—the Rockies, the Andes, the Himalayas, the Alps, and the Atlas mountains were products of the Cenozoic. This prolonged spasm of mountain building affected the shape of continents, the form of the oceans and the distribution of forests, grasslands, and deserts. During the early part of the Miocene epoch,

86

the orogenic (mountain-building) activity of the Tertiary was in full swing. In Europe the Alpine peaks were rising out of the Tethys Sea and, at the other end of this vast geological basin, the Himalayas were forming a formidable barrier between Asia and peninsular India. The rift-valley system of Africa, which extends 5,000 miles from Tanzania to Israel and the Dead Sea, was emitting its first rumbling warnings; and in the Americas the Andes, the Cordilleras, and the Cascade mountain range were being uplifted. New land surfaces were forming, and old land surfaces were being eroded and undergoing a revolution in the nature of their vegetational cover. The influence of a cooling climate which led to the withdrawal of tropical and subtropical forest belts towards the equator and the rain shadow effect, which resulted in areas of relative aridity forming in the lee of mountain ranges, contributed to a widespread expansion of grasslands. Grasslands (called prairies or steppes in temperate latitudes and savannas in the tropics) offered new evolutionary opportunities to mammals in general. The fossil record clearly demonstrates that horses, which until the Miocene were small, forest-dwelling, browsing mammals, were changing their habit, their habitus, and their habitat; they were rapidly becoming adapted to grasslands and to the grazing habit and were displaying adaptations in the teeth and limbs suited to their new environment and way of life. As we shall see, the effect of this climatic and mountain-building revolution was of considerable significance for the origins of the higher primates and man.

CHAPTER FIVE

Monkeys in motion

There is a saying among paleontologists that God made the skull and the Devil made the teeth; this is just another way of emphasizing that teeth and jaws are extremely complicated structures and that their interpretation has led to a great deal of disagreement and even bitterness among scientists.

My purpose in even mentioning this dangerous subject is to remind you that at present the theories of primate (including human) evolution are almost wholly based upon the works of the Devil. This state of affairs is, of course, inevitable when one appreciates that teeth and jaws are the commonest parts of the skeleton to be found in fossil deposits. Common, because of their durability combined, one assumes, with a certain unsavory and indigestible quality that offers little in the way of temptation to mammalian predators or scavengers which are the worst enemies of paleontologists.

When an animal dies on the African savannas it is subject to a series of insults imposed upon it in a strict system of zoological priority. First refusal goes to the initial predator, a lion, a leopard, or a cheetah which prefers, on the whole, the viscera to the flesh; the liver, the stomach, and the intestines are the first parts to be eaten. Then come the hyenas, hunting dogs, and jackals. What is left of the carcass is now the perquisite of the vultures which, with an eye on Charles Addams and their fearful reputations, pick the bones clean. Finally the bones themselves, by now thoroughly dismantled, are still a delectable food source. They are crunched up by carnivores, split open by being dropped from a great height onto a convenient rock by lammergeiers, the bearded vultures, or removed by porcupines or hyenas to their dens to be gnawed in peace for the calcium salts that they contain. Nature is very economical and the paleontologist is often left with little more than slivers of long bones, jaws and teeth, and ankle bones. This last item seems rather a curious leftover, but any anatomist will tell you that the ankle joint is one of the toughest joints in the body to disarticulate, so strongly is it held in place by thick, tendinous ligaments. The ankle bone (the talus) of *Homo habilis* from Bed I, Olduvai Gorge, only escaped being lost forever by a hair's breadth. On its upper surface a series of tooth indentations of an unknown predator can clearly be seen.

Bones of animals that die in the forest are somewhat less likely to be discovered by paleontologists of the future than those left on the savanna for, in addition to the insults offered by various sorts of predators including termites and ants, the forest corpse is a prey to the chemical composition of the soil which is highly acid and tends to decalcify any remaining fragments of bone.

It is a wonder, really, that paleontologists have anything to work on at all. Some of the best fossils are those of animals that have been preserved by accident. Animals falling into tarpits or into rivers and lakes, being inundated by volcanic ash, or being washed into deep crevices in ancient rocks are the ones most likely to be preserved.

To return to the vexed subject of teeth! Because they are so extremely slow to change in response to the pressures of natural selection, teeth are splendid structures for determining taxonomic relationships among fossil primates. Much of our knowledge of the evolution of primates is derived from studies of teeth; but for the very reason of their conservative nature, they are imperfect indicators of the evolutionary changes taking place in total bodily form. While the teeth of apes and man have tenaciously retained the overall characters that were acquired 30 million years ago by their remote hominoid ancestors, the limbs, the trunk, and the brain have undergone dramatic changes of far-reaching significance.

In this chapter we are going to look at some of these changes, particularly those concerning the limbs. Fossilized limb bones are scarce, but not so scarce as all that. Many of them are reposing unsorted and unrecognized in dusty packing cases in the basements of museums all over the world as Dr. Elwyn Simons of Yale has pointed out (1967). Scientists planning fossil-hunting expeditions might consider the possibility, in the interests of national economy, of putting away their geological maps and, instead, getting out their bus and railway timetables. In recent years, there has, in fact, been a revival of interest in primate locomotion and this has resulted in the recognition and accumulation of quite respectable collections of limb material both from the field and the museum.

Changes in environment, as we have already stressed, are associated with changes in the form of animals inhabiting these environments and it is in the locomotor system that these changes are best exemplified. The way an animal moves about depends on a great number of factors. It depends on the animal's size and it depends on the nature of the substrate on which it is moving. The precise role that an animal plays in the ecosystem of which it is a part is also relevant—whether it is predator or prey, fruit-eater, flesh-eater, browser or grazer. Nature operates by selecting those physical or behavioral characters that best fit the animal for its particular way of life. Locomotion is such a fundamental form of behavior that adaptiveness is not only reflected in the limbs and trunk but in the whole body from the top of the head to the tip of the tail.

Size plays an important part in the evolution of primate locomotion. Generally speaking the trend has been for primates to get larger as they ascend the evolutionary scale; thus New World monkeys are bigger than prosimians, catarrhine monkeys are bigger than platyrrhine monkeys, and apes are bigger than both.

Progressive increase in bodily size occurs in the lines of descent of many mammalian Orders (see Cope's Rule p. 31). There are plenty of examples amongst primates which appear to refute this generalization. For instance certain prosimians, such as the true lemurs, are considerably larger than some New World monkeys like marmosets, tamarins, and squirrel monkeys; howlers are bigger than the talapoins. Gibbons are less than a third of the weight of baboons; but these are exceptions. Generally speaking, the largest primates within each of the major taxonomic groups are the most specialized; they are the ones that have departed furthest from the ancestral pattern of structure and behavior. The existence of a trend towards increase in size is demonstrated even more positively in fossil primates than in their living descendents. From the primitive primates of late Cretaceous and early Paleocene, to the prosimian primates of the Eocene, to the undoubted forerunners of monkeys and apes of Oligocene, to the ancestral apes of the Miocene, the size progression is undeniable.

Before discussing the relationship of size and locomotion, we might pause a moment to consider why size should have been selected for, so inexorably, during primate evolution. Large size is associated with increased longevity. In fact all life periods such as the gestation period and the period of juvenile dependency and life expectancy appear to be extended as a direct effect of body size (Table 1). Increased body size is also associated with increase of brain size, increase of bodily strength, and with an absolute increase in the speed of motion. As far as the brain is concerned the larger the animal, the larger the brain and there is good evidence that, within related groups, the larger species show a greater capacity to learn and to retain the memory of things once learned, than smaller species. Perhaps this last mentioned consequence of the size trend is the most significant for primates. Clearly, however, there are also disadvantages to being big. Large animals have high energy-budgets and need large amounts of food, and in order to satisfy this demand, they must either eat bulk foods of relatively low calorific value or concentrated foods of very high calorific value. They are either herbivores or carnivores. In the former instance, their day will be spent largely in finding and consuming food with detrimental effects on the development of social and affective elements in their lives, a situation in which the gorilla finds itself. In the latter instance, food-getting and eating, being less time-consuming functions, leave a certain amount of time to spare, time which can be utilized in a variety of ways. If the carnivore happens to be a lion, it can sleep off its spare-time with luxurious abandon but, if a man, then spare-time provides a golden opportunity for the exercise of the psychosocial muscles. When man became an efficient hunter and learned about fire and cooking, a new dimension entered his life; with time to spare from the chores of mere survival he was able to develop other pursuits like painting, modeling, music, and dance. Size, however, imposes its own rules, subject to physical laws of gravity and motion; thus size and the evolution of locomotor patterns are closely correlated phenomena.

PRIMATE LOCOMOTION

The primary subdivision of animal motion is related to the medium in which it is taking place—land, sea, or air. There are no water-adapted primates nor does any primate fly or even glide. Monkeys swim naturally but they are not specially adapted for it. Man has to learn to swim; apes would also have to learn. As far as I know there is no such thing as a naturally swimming ape, nor do I know of any ape trained to swim.

Primate locomotion takes place on land which leads us to a secondary subdivision: under the ground, on the ground, or in the trees. There are no burrowing primates so we end up with two environmental categories of primate locomotion, arboreal and terrestrial.

The anatomy of primates both fossil and living indicates without doubt that the group is primarily an arboreal one. Although from the fossil record of the early Tertiary it is just possible to distinguish, amongst the early mammals, primates from non-primates, it is not until the beginning of arboreal life that the distinction can be made with any real certainty. The structural hallmarks of tree-climbing are unmistakable, but those of generalized ground living are indeterminate. In effect primates only became primates when they started to live in trees.

Terrestrial locomotion is a secondary adaptation for which, paradoxically, life in trees was pre-adaptive. The main categories of primate locomotion are listed in Table 5, column 1, and of these four major types, vertical clinging and leaping, quadrupedalism, brachiation, and bipedalism, only the latter is *exclusively* a terrestrial activity. Quadrupedalism can be deployed in either milieu but vertical clinging and brachiation are *exclusively* arboreal activities. This would seem to imply that animals in the latter two categories cannot descend to the ground. In fact they can, but in order to do so they have to adopt a completely different form of gait. On the ground vertical-clingers hop and brachiators either walk bipedally (gibbons) or with a type of gait called knuckle-walking (chimpanzee and gorilla). Unlike some arboreal specialists, the three-toed sloth for example which is virtually immobile on the ground, primates are sufficiently generalized to adapt to a new environment. Throughout evolution there has been a constant migration of primate groups between the trees and the ground. Some living species like the baboons are secondarily terrestrial, but others like the African guenons and mangabeys are secondarily arboreal having come down to the ground and subsequently reverted to a life in the trees. It is not surprising to find that so many species of primates today exploit the ground and the trees in about equal proportions.

The differences in locomotor pattern between the four primate locomotor types lie principally in the *functions of the forelimbs and the hindlimbs* (Table 5). The supporting function of the hindlimb is fairly constant throughout the Order and, therefore, it is not surprising that little variation can be seen in the anatomy of the bones and muscles of the thigh and leg. Brachiators, principally the orang-utan, use the hind-limbs in suspension; the orang's propensity for hanging by its legs is reflected, *inter alia,* by the extraordinarily hand-like appearance of its foot. Some quadrupedal primates also suspend themselves by their feet, notably South American howlers, woolly monkeys, spider monkeys, and uakaris; during this inversion they are assisted by that sublime but faintly ridiculous invention of nature, a prehensile tail.

Table 5 **Functions of the Fore- and Hindlimbs in Suspension and Support**
(+ = degree to which the limb is used; − = no use of limb in that function.)

Locomotor Category	Forelimb		Hindlimb	
	Suspension	Support	Suspension	Support
Vertical clingers and leapers (trees and ground)	−	(+)	−	+++
Quadrupeds (trees and ground)	+(+)	+++	+	+++
Brachiators (trees)	+++	+	++	++
Brachiators (ground)	−	++(+)	−	+++
Bipeds (ground)	−	−	−	+++

There is considerably more variation in the use of the forelimb than the hindlimb. Apart from bipeds and vertical clingers, the forelimb universally has a supporting function; when brachiators such as gorillas and chimpanzees are on the ground their arms contribute an equal share with the legs in supporting the body; they are, in this respect, quadrupeds. Vertical clingers hop when on the ground although, when un-hurried, they may move slowly in a quadrupedal fashion. Bipeds of course *can* use their arms for support, but because their legs are so long, quadrupedalism is an uncomfortable and ineffective way of getting about. Even human infants, who have less disproportion between arm and leg length than adult humans, crawl on their knees.

The principal function of the forelimbs of brachiators is suspension although the extent to which this is used varies quite considerably between species. For gibbons, arm-swinging is the basic pattern of locomotion; orangs and chimpanzees brachiate less often, and gorillas (when adult), not at all. Quadrupedal monkeys not infrequently use their arms to suspend the body and even cover quite considerable distances by arm-swinging. This behavioral habit led us, my colleague Peter Davis and I

(Napier and Davis, 1959) to introduce a new category of primate locomotion called semibrachiation. Although semibrachiation is an important transitional type of gait in an evolutionary sense, it is perhaps best regarded as a subtype of quadrupedalism (Table 6). The chief arm-swingers amongst the monkeys are the New World howlers, spider monkeys, and woolly monkeys. Spider monkeys have acquired a physical structure quite similar in many ways to that of the archbrachiators of the Old World, the gibbons— a further example of structural parallelism induced by similar environmental pressures. Some Old World monkeys, the langurs for example, also occasionally swing by their arms, a behavior that is reflected in the structure of their shoulder-girdles which have a shape intermediate between that of true quadrupeds and brachiators (Ashton and Oxnard,

Table 6   A Classification of Primate Locomotion*

Category	Subtype	Activity	Primate Genera
Vertical clinging and leaping		Leaping in trees and hopping on the ground	*Avahi, Galago, Hapalemur, Lepilemur, Propithecus, Indri, Tarsius*
Quadrupedalism	Slow climbing type	Cautious climbing— no leaping or branch running	*Arctocebus, Loris, Nycticebus, Perodicticus*
	Branch running and walking type	Climbing, springing, branch running and jumping	*Aotus Cacajao, Callicebus, Callimico, Callithrix, Cebuella, Cebus, Cercopithecus, Cheirogaleus, Chiropotes, Lemur, Leontideus, Phaner, Pithecia, Saguinus, Saimiri, Tupaia*
	Ground running and walking type	Climbing, ground running	*Macaca, Mandrillus, Papio, Theropithecus, Erythrocebus*
	New World semi-brachiation type	Arm-swinging with use of prehensile tail; little leaping	*Alouatta, Ateles, Brachyteles, Lagothrix*
	Old World semi-brachiation type	Arm-swinging and leaping	*Colobus, Nasalis, Presbytis, Pygathrix, Rhinopithecus, Simias*
Brachiation	True brachiation	Arm-swinging brachiation	*Hylobates, Symphalangus*
	Modified brachiation	Arm-swinging and quadrumanal climbing	*Pongo*
		Occasional brachiation; climbing; knuckle-walking on ground	*Pan, Gorilla*
Bipedalism		Standing, striding, running	*Homo*

* Modified from Napier and Napier, 1967.

94

1964). One langur, *Presbytis melalophos*, is known to travel short distances by arm-swinging alone (Bernstein, 1968).

LOCOMOTOR CATEGORIES

In the foregoing account some of the difficulties in classifying primate locomotion will have become apparent. The physical structure of primates is so generalized that it can be deployed in a variety of different ways. Giraffes do not have much alternative to being quadrupeds, and kangaroos are committed to hop for all eternity, but it is true to say that most monkeys and apes can walk and run quadrupedally, climb, leap, swing by the arms, and even indulge in brief episodes of bipedal walking. This has permitted the Order to be extremely noncommittal in terms of habitat. Taking advantage of what is offered, their generalized structure and adaptable behavior carries them through.

The following brief description of the primary locomotor categories listed in Table 6 are expressed in purely behavioral terms. Anatomy can be misleading if viewed without the illumination of the living animal and, as I have already stressed, the critical confrontation that decides the policy of natural selection is between behavior and the environment.

Vertical clinging and leaping (Figure 13a, b) is a type of arboreal behavior in which the body is at all times held vertically at rest. The body is supported against a trunk or branch by means of grasping feet and acutely bent hindlimbs; the arms and hands are applied to the trunk to reinforce the grasp of the feet which are the primary supporting agents (Plate 7). Movement is effected by the powerful frog-like extension of the hindlimbs which project the animal in a leap from one vertical support to another (as in tarsiers and indrises). Alternatively, if the leap starts from a horizontal support as sometimes occurs, the direction may be directly upwards (as in galagos) or horizontally (as in sifakas). Vertical clinging and leaping primates usually hop bipedally when moving rapidly on the ground, but assume a quadrupedal gait when moving slowly (Figure 13b; Napier and Walker, 1967).

Quadrupedalism is a type of locomotion that can take place on the ground or in the branches of trees with only minor behavioral differences between the two situations. In an arboreal milieu the opposable thumb and/or big toe may be used to grasp the branch; on the ground the palms and soles may be applied to the substrate or, alternatively, the

A

B

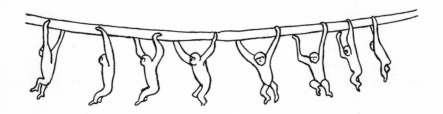

C

Figure 13. Cartoons traced from 16 mm motion picture film illustrating locomotor patterns. (A) The indris showing a series of classical vertical clinging and leaping jumps. (B) The galago *(G. crassicaudatus)* hopping bipedally on the ground. (C) Young gibbon brachiating. (A and B, courtesy of *Folia Primatologica.* A-C prepared by A. C. Walker.)

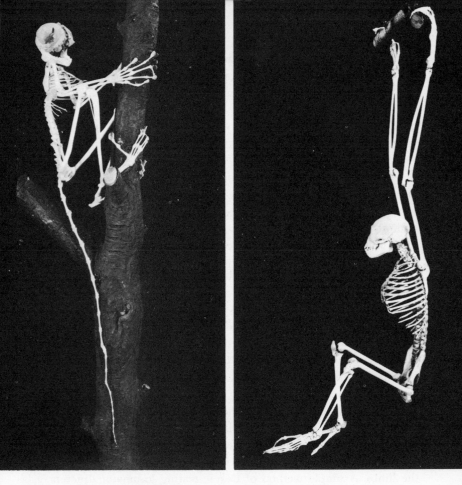

Plate 7.   Mounted skeletons of the tarsier (*Tarsius*), left, and gibbon
(Hylobates), right. (Skeletons prepared by Anatomy Department, The Royal
Free Hospital School of Medicine, London.)

undersurface of the digits only (digitigrade walking). In trees, climbing, jumping, and leaping are associated with this mode of locomotion but not necessarily. In slow moving quadrupeds (e.g., the potto), locomotion may be cautious and extremely slow with jumping, etc., quite absent. Occasional arm-swinging may also be associated with quadrupedalism especially in the New World semibrachiators. Semibrachiators (e.g., spider monkeys) when on the ground either walk bipedally or quadrupedally.

Brachiation (Figure 13c) is a form of locomotion in which the dominant component is arm-swinging by which the body, suspended from above, is propelled through space by means of a rapid alternating movement of the arms. Some brachiators (e.g., orang and chimpanzee) supplement the function of the arms by supporting activities of the legs. Brachiators on the ground either walk bipedally (gibbon), quadrupedally with hands bunched into a fist (orang), or quadrupedally on the knuckles (chimpanzee and gorilla).

Bipedalism is a form of locomotion in which the body is supported on two legs and in which progress is effected by alternating swings of the legs. The acme of bipedalism, the unique locomotor talent of man, is *striding*, which is characterized by a lengthy stance-phase in which a heel-toe propulsive action is implicit (see Chapter 8).

All four categories are linked by some aspect of the locomotor pattern which is common to each successive pair. Vertical clingers moving slowly on the ground are quadrupedal. Quadrupeds may suspend themselves, and even progress by the arms alone, in the manner of brachiating primates. Brachiators may, when on the ground, adopt bipedalism for short distances. Thus, the succession of vertical clinging, quadrupedalism, brachiation, and bipedalism forms a continuum of locomotor activity. For convenience we quite arbitrarily divide locomotion into discrete categories, in much the same way as the colors of the light spectrum are arbitrarily divided into violet, indigo, blue, green, yellow, orange, and red.

Structural adaptation to different locomotor behaviors can be discerned in most parts of the body but naturally are seen most clearly in the limbs. A simple method of demonstrating locomotor differences between species is by means of forelimb-hindlimb length ratios.

Measurements of the length of the long bones of the forelimb and hindlimb expressed as a ratio constitute a numerical expression of their relative lengths called the *intermembral index*: $\dfrac{\text{Length Humerus} + \text{Radius}}{\text{Length Femur} + \text{Tibia}}$ $\times$ 100. When these ratios are calculated for large numbers of individuals of living primates and plotted graphically (Figure 14), we see that there is a continuous range of variation from the long-legged forms (with a low index) to the long-armed forms (with a high index). This graph clearly implies a trend, but it is not obvious in which direction the trend is moving. Is it tending to produce longer arms or longer legs? It is at this point that the evolutionary approach becomes so important as a means of making sense out of a collection of data concerning living animals. We know, from evolutionary evidence, that prosimians

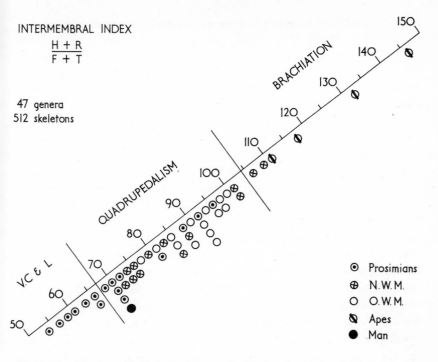

INTERMEMBRAL INDEX

$$\frac{H + R}{F + T}$$

47 genera
512 skeletons

BRACHIATION

QUADRUPEDALISM

V.C & L

⊙	Prosimians
⊕	N.W.M.
O	O.W.M.
⊘	Apes
●	Man

Figure 14.   Spectrum of values for intermembral index of primates. Based on mean of indices for each genus.

evolved before monkeys, and that monkeys evolved before apes. So the inference can be made from this graph that the trend is from the prosimian pattern towards the ape pattern—in other words from long legs and short arms towards short legs and long arms. But in order to convert inference into certainty, evidence of fossil limb-bones is needed. In fact, we have some of this evidence, as will become clear in the course of this chapter.

From the study of the behavior of living primates, as we have seen, it is possible to discern certain distinctive patterns of locomotion; these can be correlated with the intermembral index as shown in Table 7. What, you may rightly ask, is the locomotor pattern of primates with intermembral indices lying between 65-75 and 100-108? One can only answer that in the presence of an evolutionary trend *formes de passage* must be expected. There are a number of primates that show a style of locomotion that is intermediate between locomotor categories. Some Madagascan lemurs have indices which fall between 65-75, their gait is a combination

Table 7 Correlation Between Locomotor Pattern and Limb Proportions

Locomotor Pattern	Intermembral Index (percent)	Limb Proportions
Vertical clinging and leaping	50–65	Short arms, long legs
Quadrupedalism	75–100	Equal or subequal limbs
Brachiation	108–150	Long arms, short legs

of vertical clinging and quadrupedalism. *Lemur catta* can often be observed leaping from one vertical support to another in the manner of vertical clingers, but more commonly the gait of lemurs is quadrupedal and is generally classified as such. Having very short arms and long legs, however, *L. catta* is not particularly well adapted to a quadrupedal gait. Had it been in competition with better adapted quadrupedal monkeys, *L. catta* probably would not have survived, but being in the "refuge" area of Madagascar where there are no monkey competitors, it has managed to get by. The most notable primate whose intermembral index falls between 100-108 is *Ateles,* the South American spider monkey. Again, this animal shows a very "intermediate" sort of locomotion. Spider monkeys are partly quadrupedal and partly brachiating, as much time being spent swinging by arms (and tail) as walking on all four limbs. *Ateles* occupies the brachiating niche of South American forests. If it had, by chance, evolved in the Old World it might not have survived the competition of more efficient brachiators like the gibbon.

There are several other indices which provide information on locomotion, such as the brachial index (the relative length of the forearm to the upper arm), the crural index (the relative length of the lower leg to the thigh), the humero-femoral index (the relative length of the upper arms to the thigh), and innumerable other indices derived from the proportions of the hand and the foot that need not concern us here; none of these are as suitable or as informative as the intermembral index. In addition, much information that is relevant to locomotion can be obtained from various other parts of the skeleton, the skull and vertebral column, for instance.

LOCOMOTOR MILESTONES TO HUMAN EVOLUTION

To study the origins of primate locomotion is also to study primate evolution. The ability to move from place to place is fundamental to the three basic behavioral patterns of primate life—feeding, escape from predators, and reproduction. It is little wonder then that natural selection has acted so strongly on the locomotor system and produced such a wide range of adaptive types amongst primates. The manner in which monkeys move in terms of speed and agility and the environment in which they move almost wholly dictates the shape of their bodies and the patterns of their behavior. Dr. S. L. Washburn of the Berkeley campus of the University of California has written (1950) that changes in locomotor pattern provide the principal milestones along the evolutionary pathway from the Eocene to the Miocene. For the rest of this chapter we shall be tracing the evolution of non-human primates by studying some of Washburn's locomotor milestones. The first of these bears the inscription:—

Sixty Million Years to Man The earliest primates of the Paleocene were small creatures, little bigger than a white rat and not at all dissimilar to them in general appearance. Long bodies, short limbs, elongated, quivering snouts, and clawed hands and feet were the hallmarks of the primordial primates. Not for nothing have they been called the rats of the Paleocene! True rodents were in fact late comers to the mammalian scene (Figure 15). Since then rodents have been the chief competitors of primates and the most successful mammalian group next to man himself. At the beginning of the Paleocene the roll call of mammals was not a very impressive affair. There were insectivores (which seem to fit the bill as representatives of the most primitive mammalian stock); there were creodonts or primitive carnivores, and condylarths, the forerunners of ungulates, or hoofed mammals. Paleontologists have the utmost difficulty in assigning early Paleocene fossils to their appropriate Orders simply because at this stage primitive mammals were so very similar. The principal elements of distinctiveness are the teeth and the structure of the bony ear.

The adaptive radiation that produced the incredible diversity of mammals—the phenomenon that makes zoos such interesting places to visit—had not really begun when the primates came into being. Both in the evolutionary sense and in the sense of their ascendency in the world of animals, primates are primary.

The fossil evidence of early Paleocene primates suggests that while they may have been forest-dwellers, they had not yet acquired the knack of tree-climbing. None of the adaptations of skull or limbs by which we recognize primates today were present at this time. A long snout, a small

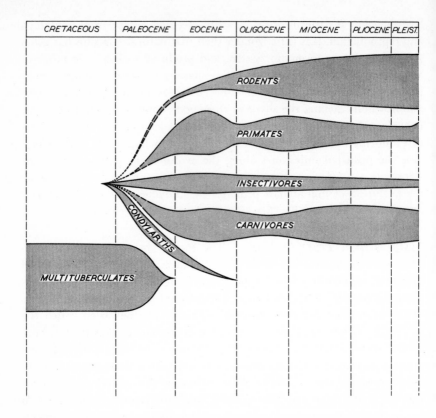

Figure 15. Evolution of basic mammalian orders during the Tertiary period.
The width of the areas reflects frequency of occurrence of the Orders in
fossil deposits throughout various geological epochs. (After Alfred S. Romer,
Vertebrate Paleontology. 2nd edition. University of Chicago Press, 1945.)

brain, eyes directed as much sideways as forwards, a long body and short
clawed limbs suggest that they had the build, and probably the gait, of
squirrels or other rodents. These characters would have fitted them to live
at all levels of the forest or forest-floor much as squirrels and treeshrews
do today. Tree-dwellers, they may have been, but they were not specialized
tree-climbers. The teeth of the Cretaceous ancestors of the primates had
sharp high pointed cusps ideal for cracking the hard chitinous exo-
skeletons of insects which probably constituted the main part of their
diet. During the Paleocene, cusps of primate teeth became less tall and
sharp and, as Frederick S. Szalay of the American Museum of Natural
History suggests (1968), this even may have been correlated with the

adoption of a fruit-eating habit. This dietary shift must not be taken too literally; all primates from treeshrews to man include insects in their diet which provide a major source of protein. The addition of a bulky food like fruit was probably associated in evolution with an increase in body size. Insects are a laborious way of getting enough to eat unless one happens to be very small and easily satisfied, or has the extensile, sticky tongue of an anteater. Such a change may also have necessitated a shift from a nocturnal to a diurnal habit.

Increase in size and the gradual addition of fruit to the diet of early primates would have had repercussions on the nature of the habitat and, thus, on the pattern of locomotion. Fruit-eating is more easily accomplished in trees and it may well have been this growing preference amongst primates which initiated the tree-climbing habit. Once a preference such as fruit-eating is established and proves successful, nature operates to enhance the effectiveness of the behavior by means of selection of suitable physical adaptations should they crop up in the genotype. In this case the most germane adaptations would have been those that facilitated climbing, balancing, and jumping in trees. Other suggestions to account for the adoption of arboreal life by the earliest primates have been mooted at various times. For instance, Frederick Barth suggested (1950) that ecological pressure by the rapidly expanding rodent Order drove the primates from the forest floor habitat into the relatively unpopulated (by mammals anyway) domain of the forest canopy.

It is noteworthy that in phylogeny prehensility of the foot preceded the prehensility of the hand. Vertical clingers, like the tarsier, or small squirrel-like quadrupeds, such as the marmosets, *by remaining small* have never set into motion the string of anatomical consequences that accompany increase in size. Prehensility of the hand is one of them. A large arboreal animal without prehensile hands is in no great danger of falling, providing the branch on which it is moving is firm and wide, but these conditions severely restrict its feeding range. In trees with broad spreading crowns, leaves and fruit are most abundant towards the periphery, but it is in just this zone that branches are becoming slender and springy. A large animal on a slender, swaying branch is in a hazardous position without some means of holding on. If it cannot *balance* on the branch there is no alternative but to *suspend* itself from the branch. Suspension by the feet alone is one possible solution but to possess the faculty of suspension by all four limbs is infinitely more effective. Feeding is not the only factor that prompts a primate to utilize the slender, flexible branches of the edge of the crown; they are much favored as sleeping sites being relatively inaccessible to large nocturnal, non-prehensile predators such as leopards and jaguars.

Virginia Avis (1962) was the first scientist to draw attention to the close relationship between feeding habits, locomotor patterns, and the microhabitat. To this triumvirate, I would add the important consideration of increasing body size. Increase in body-size, it seems to me, was the principal evolutionary trend that promoted many of the arboreal adaptations of primates.

EVOLUTION OF PROSIMIAN GRADE *Arboreal specializations* are not seen in the fossil record of primates until the middle Eocene. At this time the mushroom cloud of the primate explosion was expanding rapidly and a variety of families had come into being. These families included the probable ancestors of the lemurs and lorises (Adapidae), the tarsiers (Tarsiidae), and the monkeys and apes (Omomyidae). Here was the first great parting of the ways which separated higher primates from lower primates (Figure 4). The Adapidae contains two North American genera, *Notharctus* and *Smilodectes,* well known from a number of incomplete skeletons. *Smilodectes* had a shortish face, forwardly rotated orbits with the orbital ring fully closed in front, but still open behind (Plate 8). The hindlimbs were long and the forelimbs relatively short with an intermembral index of 63. The big toe was very large and strong and, separated from the other toes, was capable of providing a prehensile grip; the thumb was somewhat divergent, but not yet opposable. The digits were topped by flattish nails. The teeth were very similar to those of present-day lemurs except that the primitive mammalian pattern of four premolars had not yet been lost.

Smilodectes was an inhabitant of North American forests, but other species of the same family are fairly common in Eocene fossil deposits of Europe; it is among the European Adapidae, notably the genus *Adapis,* that ancestors of modern prosimians are to be found. After the middle Eocene, zoogeographical evidence (see Chapter 4) makes it rather unlikely that North American primates could have migrated successfully into Europe or Asia.

The second family evolving during the middle Eocene was the Omomyidae. Representatives are found both in North America and in Eurasia; it is probably the key family for tracing the roots of mankind. According to Dr. Elwyn Simons, no other Eocene family has a greater

Plate 8.　Mounted exhibit of *Smilodectes* at the National Museum of
Natural History, Smithsonian Institution, Washington, D.C. (Smithsonian
photograph no. 2129.)

claim to be ancestral to apes, monkeys, and man. However, the missing link or links between omomyids of the Eocene and the anthropoids of the Oligocene has not yet been found. Unfortunately no limb bones of Omomyidae have been preserved, so there is little that can be said about their locomotion, except inferentially from the skull structure, which suggests an upright posture of the trunk similar to that of vertical clingers and leapers.

Finally, the third important European Eocene family should be mentioned, the Tarsiidae, if only for the reason that, for once, we can be dogmatic and say here is an Eocene family that we know with certainty has survived to modern times. The fossil remains of the hindlimb of *Necrolemur*, a tarsioid from the middle Eocene, indicate adaptations for vertical clinging and leaping in the femur, the tibia, and the heel bone, which is markedly elongated, just exactly as we see these characters today in the living descendent of this family, the tarsier (*Tarsius*). To use a much overworked term—the tarsier is a living fossil, a cliché for which no apology is needed because the late Eocene tarsioids are almost identical in skeletal form with the living species (Plate 7). These three families, then, the adapids, the omomyids, and the tarsiids are, as we see it at the moment, the stocks from which all living primates (treeshrews excepted) arose.

The identification of vertical clinging and leaping as the possible ancestral mode of locomotion is of considerable significance to our understanding of primate evolution. It offers a possible explanation for the characteristic long-legged morphology of the living quadrupedal semi-brachiators and for the long-legged morphology of man. We shall be returning to this idea in a later chapter; suffice it to say here that the vertical clinging and leaping habit of Eocene forerunners of the anthropoids is one of the roots of mankind that this book is committed to explore. Now the second milestone is coming up, and it reads:—

Thirty Million Years to Man   By the middle of the Eocene, the arboreal habit of primates was clearly established. Prosimians, such as the omomyids, were showing a number of characters which seem to foreshadow the anthropoid condition. For instance, the number of teeth in the omomyid dentition was identical with that of present-day platyrrhine monkeys, the muzzle was foreshortened and the eyes were rotated well to the front of the face. We can infer from other Eocene families such as the closely related Adapidae that the big toe was opposable and the thumb, though not yet as opposable as the big toe, was capable of diverging from the rest of the fingers, thus giving the hand a considerable degree of prehensile ability. The development of grasping power in the hands of the earliest monkey-like prosimians may have been related to an increase in size, to the evolution of a new arboreal niche, and a new feeding habit.

EVOLUTION OF NEW WORLD MONKEY GRADE The earliest South American primates are known from the Miocene, and the interesting fact about them is that they appear almost indistinguishable from certain South American species. One can conclude in other words that New World monkeys had reached their evolutionary "ceiling" by Miocene times and that, since then, relatively little adaptive change has occurred in the platyrrhine stock. Nevertheless it is likely that the long limbs and the brachiating habits of the spider, woolly, and howler monkeys evolved somewhat later, just as parallel brachiating specializations did in the gibbons, for instance, amongst the catarrhine primates.

EVOLUTION OF OLD WORLD MONKEY GRADE The first substantial evidence of catarrhines in the fossil record comes from the Fayum region of Egypt where fossils of an Oligocene age have been found, etched out of the unconsolidated sand by the action of wind erosion. The Oligocene specimens of the Fayum are not excavated, not dynamited, but are simply picked up from the surface after the wind has blown off the concealing sands of the desert. Elwyn Simons, leader of the Yale expedition, which has made so many discoveries in this area in recent years, has described fossil hunting in this region as "arm-chair paleontology." I hesitate to accept this somewhat "British" understatement.

A number of different genera have been identified in the Fayum deposits including *Apidium, Propliopithecus, Parapithecus, Aegyptopithecus* (Plate 9), and *Aeolopithecus*. Among this group the fossil remains of which, incidentally, could all be placed in a shoe box with room to spare, are the possible ancestors of Old World monkeys (*Parapithecus* and *Apidium*), gibbons (*Aeolopithecus*), apes (*Aegyptopithecus*), and man (*Propliopithecus*). Unfortunately, aside from a few fragments of postcranial bones, we know little about the limbs of Oligocene catarrhines and therefore little about their locomotion. However, we do know some-

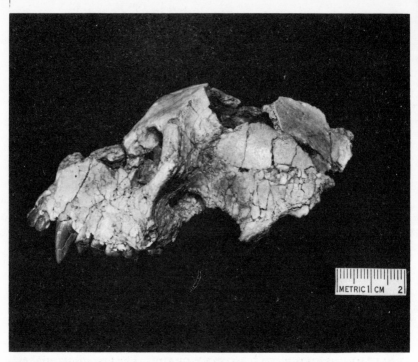

Plate 9. Skull of *Aegyptopithecus*. (Photograph by Elwyn Simons.)

thing of the probable Eocene forerunners of the early catarrhines and, as you will see, we know quite a lot more about their Miocene descendants. So, to do a little educated guessing about the intervening stage is not too heinous a crime against scientific principles. The Omomyidae of the Eocene, we believe, adapted to life in the canopy of the trees and moved about by the specialized leaping adaptations that have already been described. They were fruit eaters but their diet probably was reinforced with animal protein derived from insects, small birds, and reptiles. Proportionally their legs were considerably longer than their arms—the intermembral index was 60-65; the hands were prehensile although the thumb was not yet fully opposable. The big toe, however, was opposable. The muzzle was shortish, the eyes were forwardly rotated, facilitating stereoscopic vision, and the brain was relatively large.

The catarrhine primates of the early Miocene were arboreal and, as we can infer from the proportions of the limbs, they moved about much as certain arboreal monkeys, such as the colobines, do today. Proportionately, the hindlimbs were 10-15 percent longer than the forelimbs with an inter-

108

membral index of 80-90, and they had not yet lost their tails. The early Miocene primates were also considerably larger than their Eocene fore-runners.

With these two sighting shots, the catarrhine primates of the Oligo-cene must inevitably turn out to be small to medium-sized creatures, omnivorous in diet, intermediate in locomotor function between the extremes of vertical clinging and leaping, and of quadrupedalism. Anatom-ically the hindlimbs would have been 20-30 percent longer than the fore-limbs (intermembral indices 70-80). Their thumbs would have been divergent but not yet opposable. The brain case would still have been rather small compared to the facial skeleton, but relatively longer than in most mammals, the eyes frontally directed, and the external auditory canals rather short; the teeth (as we know to be the case from fossil evidence) bore the seal of catarrhines.

In effect, apart from the teeth, the hypothetical form we have just described resembles a New World monkey—a titi or a douroucouli (inter-membral indices of 74 and 74 respectively). In fact, *Aotus* (the dourou-couli) retains some of the traits of vertical clingers and is known, occa-sionally, to adopt a vertical clinging posture on the trunk of trees. During the early and middle Oligocene, then, the "model" for early catarrhine evolution is to be found among the more primitive living genera of the New World monkeys. This is not too surprising a conclusion when one realizes that the platyrrhines and catarrhines evolved from closely related omomyid stocks in the middle-to-late Eocene. In view of this it could be expected that early catarrhine primates would share many anatomical features with their early platyrrhine cousins. Even in the early Miocene there are still traces of New World monkey features. For instance in the European genus *Pliopithecus* the external bony tube of the ear was short as in New World monkeys; and in the East African *Proconsul africanus* the thumb was pseudo-opposable (see p. 182), a typical feature of platyrrhines. We have only to look at New and Old World monkeys today (40-45 million years *after* the postulated point of separation of the two stocks) to see that both are built, virtually, on the same plan. Is it not probable, a mere 10 million or so years after the two groups went their different ways, that their descendants would be even more alike then than they are today?

During the Oligocene, therefore, we can say that the New World monkey grade was probably established and that platyrrhine and catar-rhine primates were gradually moving away from a common vertical cling-ing ancestry towards a more proficient way of utilizing the food resources available to them by means of quadrupedalism. It was probably about this time that leaves became an important new food resource; a leaf-eating specialization would have substantially increased the range of leaf eaters compared with fruit eaters, permitting them to thrive in mixed deciduous forests where fruiting is a seasonal event but where leaves are edible all year round. By the Miocene the quadrupedal way of life was well established. Fossil primates are known from this period in Europe and in East Africa.

Until the late Miocene when cooler conditions were beginning to prevail, central Europe was still subtropical. In many regions grasslands were replacing the forests and some Old World monkey stocks were taking advantage of the availability of a new habitat by adopting a ground-living way of life. These monkeys, the cercopithecines (subfamily Cercopithecinae), are likely to have comprised the ancestral stock from which baboons and macaques are descended. Not all monkeys took to the ground, the more "conservative" members of the Cercopithecidae, the colobines (subfamily Colobinae) which were better adapted for exploiting the forest food resources perhaps, pursued the even tenor of their ways in tropical and subtropical forests consolidating the advantageous position that leaf-eating adaptations had already given them. The "conservative" colobines comprise the stock from which all modern langurs, of the genus *Presbytis* and other related genera (such as *Colobus*) are derived.

Ground-living led to a modification of the quadrupedal pattern of locomotion. Selection pressures would have been operating in favor of adaptations that improved the fitness of ground-living monkeys, which by then were in competition with a host of other mammals—the horses, rhinoceroses, giant pigs, deer, camels, llamas, and ancestral giraffes—that thronged the Miocene savannas.

By the end of the Miocene the quadrupedal grade was well established with already two distinct sorts of quadrupedal monkeys in existence. Firstly, an arboreal group with an intermembral index in the region 75-85 indicating long hindlimbs and rather short forelimbs, and a body form well adapted for leaping; and, secondly, a ground-living group with intermembral indices of 85-100 signifying that the fore- and hindlimbs were equal or subequal in length. *Mesopithecus,* a fossil cercopithecine of the upper Miocene of Greece had an intermembral index of 88, a clear indication that the cercopithecines of the late Miocene were just about getting into their terrestrial stride. The next, and last, milestone reads:—

Twenty Million Years to Man   Whereas monkeys of the Miocene looked much like monkeys of today, the apes of the Miocene most decidedly did not look like modern apes—they looked like monkeys. This sounds thoroughly paradoxical but it is really quite simple. Monkeys and apes evolved from the same stock (Figure 4) and for many hundreds of generations after their separation, they would have continued to look alike because, as far as we know, both groups continued to occupy the same sort of environment. In terms of locomotion, their problems would have been much the same so that the general body build could not be expected to have been so very different. We know, for instance, that in the Miocene, *Pliopithecus,* a gibbon-like ape, possessed a long tail and limbs of monkey-like proportions.

Evolution of an organism is not simply a harmonious and integrated transmogrification but a conversion that proceeds irregularly in bits and pieces and stops and starts. At any given moment in time an organism is a patchwork of conservative and progressive features. It is no matter of surprise, therefore, that transitional animals should be mosaic-like in their structure—here a bit of monkey, there a bit of ape. As it happens locomotor changes, the subject of this chapter, tend to be all pervasive, producing adaptations that affect the appearance of the body as a whole. On the other hand, changes resulting from new food habits may only show their effects in the structure of the teeth and in the form of the salivary glands and digestive apparatus.

EVOLUTION OF THE APE GRADE   It is quite possible that the principal factor which led to the separation of Old World monkeys and apes, was a specialization in diet. Apes, today, are substantially fruit-eaters while monkeys are either leaf eaters (colobines) or omnivorous (cercopithecines). Following the initial behavioral preference for one sort of food or another, natural selection would have produced quite rapid changes in the nature of the dentition. The changes needed to convert an ape type of molar pattern, with its obliquely arranged crests, to a monkey type in which the crests between the cusps are oriented transversely (bilophodont condition), are not very profound. During the Oligocene and early Miocene it is highly likely that the *only* difference between the ancestors of present-day monkeys and present-day apes was the form of the teeth; even apes were still in the "monkey grade" of primate evolution. Dr. S. L. Washburn has referred to these Miocene apes as "dental apes."

The ape grade of evolution, when fossil apes started to look like modern apes, was long delayed and apart from one small scrap of evidence to the contrary it appears to have been initiated some 12 million years ago. Ironically, the first recognizable "ape" in the fossil record was not a true ape at all, but as we are concerned in this chapter with the evolution of locomotor patterns rather than of taxonomic groups, *Oreopithecus* is as good an example of a brachiator as anything else. *Oreopithecus* has been known since 1872, but it was not until 1958 that a complete, though grossly distorted, specimen was discovered plastered to the roof of a tunnel in a brown-coal mine in Baccinello, Italy. The specimen, flattened by heat and pressure, was carefully removed *en bloc* from the surrounding soft coal by its discoverer, Dr. Johannes Hurzeler and transported to Basel, Switzerland, where it is still being studied in detail. A preliminary appraisal of the skeleton, carried out by William L. Straus, Jr., of the Johns Hopkins Medical School has given us a very good idea of the anatomy of the "abominable coalman" as it has been called. *Oreopithecus* had rather a short face reminiscent of *Proconsul africanus* and a brain cavity as large as that of a chimpanzee. Its teeth, which are very strange and quite different in form from those of monkeys and apes and man, are the principal reason why *Oreopithecus* has been banished from the Hominidae, the Pongidae, and the Cercopithecidae, and left to mull over the unfairness of life in a family all its own. But it is the vertebral column and limbs of *Oreopithecus* that concern us most.

The character that first and foremost suggests that we are dealing with the ape grade is the intermembral index (119) which is comparable to that of a gorilla (Table 8). Long arms and relatively short legs are the hallmark of the brachiator. Brachiation as we have already discussed is the characteristic locomotor pattern of the ape-grade. Other brachiator-like features of *Oreopithecus* are a broad chest, a short lumbar segment to the vertebral column, a pelvis with expanded iliac blades and a broad sacral area. These last two features also suggest that *Oreopithecus* was capable of a certain amount of bipedalism. Finally, the hands with their long rather curved fingers are very similar to the hands of brachiating apes. All in all, we can say that with *Oreopithecus* the ape grade had arrived.

EVOLUTION OF THE HUMAN GRADE   By the Pliocene three of the four locomotor patterns, vertical clinging, quadrupedalism, and brachiation,

Table 8   Intermembral Indices of Some Primates (N=Number of Specimens)

	N	Mean		N	Mean
Alouatta	18	99	Lepilemur	11	61
Aotus	10	74	Macaca	31	92
Ateles	16	106	Nasalis	23	93
Avahi	7	56	Nycticebus	8	91
Callicebus	7	74	Pan	13	108
Cebus	12	81	Papio	18	95
Cercocebus	13	82	Perodicticus	9	87
Cercopithecus	42	82	Pongo	31	145
Colobus	29	80	Presbytis	88	78
Erythrocebus	7	91	Propithecus	10	62
Gorilla	17	116	Saguinus	20	77
Homo	33	69	Saimiri	14	79
Hylobates	19	131	Tarsius	16	56
Indri	8	64	Tupaia	16	73
Lemur	22	70			

were in evidence. What of the fourth—what of bidepalism? The origins of this locomotor pattern are discussed at length in Chapter 8; suffice it to say at this point that there is no certain evidence that bipedalism evolved before the Pleistocene; however this is to take a very pedantic attitude. What is meant is that the first fossils that *demonstrate* the existence of bipedalism occur in the lower Pleistocene, but common sense tells one that these Pleistocene bipedalists *must* have had ancestors. Bipedalism is a complicated business and it could certainly not have evolved overnight. Without any shadow of doubt there were bipedal forms existing in the Pliocene. Elwyn Simons and David Pilbeam have put up a good argument that *Ramapithecus* of the late Miocene was bipedal (see p. 192). Scientifically their arguments, based as they are on inference, are simply to be regarded as an interesting hypothesis. Commonsense-wise they are probably right. It seems rather sad that "science" and "commonsense" can be at such variance and that the conventions of science impose an obligation to deny what is probable.

We are not so much concerned in this chapter with the evolution of locomotion in any particular primate stock, such as the apes or man, as we are with the *trends of locomotor change* seen throughout primate evolution as a whole. If we stand back a little and are not too distracted by irrelevant detail, we can see evidence of the successive development of four different sorts of locomotion: vertical clinging and leaping, quadrupedalism, brachiation, and bipedalism (Table 9). All living representatives of the primates can be placed in one or the other of these broad locomotor categories. We can also recognize many intermediate forms whose evolution has, as it were, been frozen in a transitional stage. Evolutionary freezing, or stasis, is not in itself a bad thing; in fact it is quite the opposite. Evolutionary success may be assessed on the basis of survival over a long period of time; thus such primates as the lemurs are in these terms successful because early in their evolution they reached a

Table 9 **Evolutionary Grades and Corresponding Locomotor Patterns**

Grade	Locomotor Pattern	Epoch
Prosimian	Vertical clinging and leaping	Eocene
Monkey	Quadrupedalism	Miocene
Ape	Brachiation	Pliocene
Human	Striding bipedalism	Pleistocene

state of equilibrium with the environment that has never been lost. They reached an adaptive plateau and have been humming along nicely ever since. Not only are transitional forms, like the true lemurs, in this tricky position but so are a number of other primate species whose locomotor function is transitional. But for all that, they are respectable primates that are not to be looked upon as outsiders who have not had the simple decency to conform to a man-made concept of locomotion.

The principal locomotor trend throughout primate evolution has been from a *hindlimb-dominated gait to a forelimb-dominated one*—from vertical clinging to brachiation (Table 7), from the tarsier type to the gibbon type (Plate 7). Among other things, this trend has tended to emphasize the persistence of the upright posture throughout the course of primate evolution and this, perhaps above all, has paved the way for the emergence of man. Man is a product of three intimately linked possessions—his upright posture, his manipulative hands, and his large brain. Without his hands he would not have a large brain and without his upright posture he would not have skillful hands.

Spinning off from the four primary locomotor types are several important subgrades, which can be regarded as side specializations, marginal to the main trend of evolution. For instance, the pottos and the lorises, according to one primatologist, Dr. Michael Blake, are specialized offshoots of the vertical clinging or prosimian grade. Baboons and their close relatives, the mandrills, the drills, and the macaques, are arboreal quadrupeds which have taken up life on the ground and have become structurally adapted to this milieu. Gorillas and chimpanzees are brachiators which have secondarily become adapted for knuckle-walking locomotion on the ground. Man himself is a spin-off from the main locomotor trend, although at just what point he did so is a matter for further discussion (see Chapter 8).

CHAPTER SIX

Apes in the attic

The following letter, which appeared in the correspondence columns of the London *Times* (5 November 1958), produced only a handful of material results but a shower of facetious and world-wide comment. As a fall-out of what was a perfectly serious inquiry, the rubric "Apes in the Attic" has given the writer excellent service as a lecture title for some years.

APES IN THE ATTIC

TO THE EDITOR OF THE TIMES

Sir,—May I on behalf of the Anatomical Society of Great Britain and Ireland appeal through your columns to those of your readers who have apes stored in their attics? This learned society is undertaking the task of compiling a catalogue of all existing skeletons of the four anthropoid apes (gorilla, chimpanzee, orang-utan, and gibbon) in Great Britain and Ireland, for purposes of scientific study. As this material is of great scientific value the society would be extremely grateful if private collectors prepared to cooperate in such a project would communicate with me at this address.

Yours faithfully,
JOHN NAPIER, Reader in Anatomy.
Royal Free Hospital School of Medicine,
8, Hunter Street, W.C.1, Nov. 4.

It is the apes in our attic and the skeletons in our cupboards that this book is all about. The word "ape" in common parlance is used to describe anything from a marmoset to a gorilla. It is a far older word than monkey and has been in use for the best part of 2000 years. The ancients knew nothing of true apes but were very familiar with monkeys which figure in Egyptian paintings and sculptures of pre-dynastic times. In modern scientific jargon the term ape is reserved for the tailless Old World primates—the true apes (gorilla, chimpanzee, orang, and gibbon). The origin of the word monkey is something of an etymological mystery. One ingenious, though purely hypothetical, explanation derives it from the Greek word *pithecus* meaning an ape. When written in classical Greek characters, the genitive case of *pithecus* looks like this:

$$\pi\iota\theta\eta\kappa o\upsilon$$

One supposes that the word could easily have been read at its face value as "monkey" by a student unfamiliar with the Greek alphabet. While speculative to a degree, this explanation is quite as convincing as the derivation posited by the Oxford English Dictionary!

At any rate whatever the philologists discover in the future, in *our* usage an ape is an ape and a monkey is a monkey.

WHAT ARE THE APES?

The apes constitute two closely related families of the suborder Anthropoidea: Hylobatidae and Pongidae. Zoologically they have equal status with the platyrrhine families (Cebidae and Callitrichidae), the catarrhine family (Cercopithecidae) and the human family (Hominidae). The close relationship of apes and man is reflected by their common grouping in the superfamily Hominoidea:[10]

Superfamily Hominoidea Simpson, 1931
    Family Hylobatidae Blyth, 1875
        Genus *Hylobates* Illiger, 1811
            Symphalangus Gloger, 1841

---

[10] In formal usage, such as the partial taxonomic list given here, the scientific name of the family, genus, or species in zoology is followed by the name of the author who first coined it and the date of the work in which it first appeared. It will be noted here—and throughout the text—that genus and species names are italicized. This is a universal practice.

Family Pongidae Elliot, 1913
  Genus *Pongo* Lacépède, 1799
    Pan Oken, 1816
      Gorilla I. Geoffroy, 1852
Family Hominidae Gray, 1825
  Genus *Homo* Linnaeus, 1758

As man is the principal subject of the last three chapters, and gibbons and orangs have already been considered in Chapter 3, here we shall only be considering the structure, behavior, and evolutionary history of the two African apes, the chimpanzee and the gorilla, which form man's closest zoological link with his past.

Anatomical, physiological, and biochemical studies have shown that gorillas and chimpanzees are phylogenetically closer to man than are gibbons and orangs. In fact so closely related do they appear in the nature of certain immunological reactions of the blood proteins that some biochemists (e.g., Goodman, 1963) are prepared to include the African apes and man in a single family. This, however, is the unorthodox view. Equally unorthodox and inconsistent with the evidence of paleontology is the view, for which Dr. S. L. Washburn of Berkeley is the leading protagonist, that the human line evolved from the ape line some 5 million years ago. The knuckle-walking habit of African apes when on the ground represents, according to Washburn, the antecedent stage to human bipedalism. Recently the studies of Sarich (1968) have provided considerable support for Washburn's view. It appears that body proteins, reflecting as they do the structure of the genetic material itself, provide a means of studying within a living species the record of its own evolution. It would seem that evolutionary information can be derived from a comparison of the speed and sensitivity of the immunological response between a serum protein, such as albumin, taken from a series of living primate species, and antisera prepared in rabbits. The difference in amino acid sequences between albumins of any two species of primates, thus tested, is expressed quantitatively as "immunological distance." The immunological distance, for instance, between man and an Old World monkey is 36, that between man and a New World monkey 58, and between man and a tarsier 114. These distances correspond to the generally accepted taxonomic placement of the four superfamilies; this permits the assumption that there is a correlation between the length of *time* the species have been separated in evolution and the amount of *difference* that has accumulated in their albumins since the separation. Clearly a further assumption must be made before such data can be used as characters of phyletic signifi-cance: that is, that mutation rates producing changes in the chemical structure of proteins have proceeded at a uniform rate during evolution. This assumption has recently been subjected to strong criticism (Read and Lestrel, 1970). Sarich's "albumin time-scale" has been applied to the homi-noids, and from his studies of albumin from the great apes and man he concludes that man diverged from the gorilla and chimpanzee stem some 4-5 million years ago at the end of the Pliocene.

Unfortunately this apparently simple method of solving the problem of the date of man's separation from the African apes is not in accord with the fossil evidence. As will be discussed later in this chapter there is evidence that even in the early Miocene a gorilla-like pongid was already in existence; and that by the late Miocene the man-like forms *Ramapithe-cus* and *Kenyapithecus* had already appeared on the scene, antedating Sarich's proposed separation point by some 10 million years. It is difficult to reconcile the contradictions of immunological and paleontological in-formation in terms of time unless it can be assumed that apes and humans evolved more than once and that the Miocene representatives of these two stocks became extinct leaving only a heterogeneous group of hominoids which, late in the Pliocene repeated the experiment successfully. This I suppose is a remote possibility. The more generally accepted view of ape and human relationships places the point of separation in the neighbor-hood of 20-25 million years ago. The dendrogram shown in Figure 17 approximates the current consensus view to which I subscribe.

Although chimpanzees and gorillas are closely enough related to each other to persuade the eminent zoologist and paleontologist, George Gay-lord Simpson of Harvard, that they are simply two species within a single genus, they differ significantly in many aspects of their structure, ecology, and behavior (see Table 10). Simpson's view is by no means universally accepted among primate biologists.

Chimpanzees are forest-living and essentially gregarious animals that live in large bands which are constantly changing in composition. This form of "open" society in which any sense of territoriality seems to be lacking is probably related to the exigencies of food-finding. Fruit, their principal diet, is widely dispersed and the location of fruit trees changes annually and demands a wide home range and a ready tolerance of other groups at feeding sites. Codes of social interaction within the band, although they undoubtedly exist, appear to lack the intensity and rigid-ity with which they are enforced in savanna-living baboon troops. Male-

Table 10 **Comparison of Ecology and Behavior in the Four Apes***

	Chimpanzee	Gorilla	Orang-utan	Gibbon
APE ECOLOGY				
Habitat	Tropical rain forest, Africa Montane forest Tropical woodland	Tropical rain forest, Africa Montane forest including bamboo zone	Tropical rain forest, S.E. Asia	Tropical rain forest, S.E. Asia
Niche	Arboreal and terrestrial	Terrestrial and limited arboreal	Arboreal	Arboreal
Diet	Fruit Pith, leaves, bark Insects Meat	Pith, vines. leaves Fruit	Fruit Bark, leaves Insects, birds' eggs	Fruit Pith, leaves Insects, birds' eggs, nestlings
APE BEHAVIOR				
Reproductive unit	Multi-male groups	Multi-male groups	? Family group	Family group
Group size	Variable. May be as large as 50	Variable. 5-30	2-4	2-4
Nesting behavior	Day and night nests in trees	Night nests in trees or on ground	Night nests in trees	No nests
Sexual dimorphism	Moderate	Very marked	Very marked	None

*Modified from Reynolds, 1967.

dominance behavior is apparent in chimpanzee bands but its application in maintaining the social status quo, in ensuring harmony among troop members and in sexual selection, is not intrusive. The difference in behavior in chimpanzee bands and baboon troops, which operate under "closed" society principles, is analogous to the human situation of war and peace. Baboons are always at war with their environment so that territoriality, intratroop discipline, and predictability of behavior, essential for survival, is of high selective value. Chimpanzees on the other hand are at peace; chimps are the flower children of the apes, irresponsible and given to bursts of noisy, but apparently happy activity; food is plentiful all year round, predators are few and the "lazy, hazy, crazy days of summer" last all the year long. The essential contrast between baboon and chimpanzee societies is clearly expressed by the antithetic implications of the collective nouns, troops and bands, commonly used to describe their respective aggregations.

Chimpanzees are quite a prolific species, and are found in most of the rain forest areas of Africa from the extreme west to the Rift Valley in the east. As a species, they are in no immediate danger of extinction although in some regions, such as Sierra Leone, local populations come under considerable pressure from local native hunters, many of whom are equipped with modern firearms and kill adult chimpanzees for food and capture the young for export. In many areas local populations of chimpanzees have disappeared in the last decade or so. During the capture of an infant the mother is almost inevitably killed; and so high is the infant mortality rate in captivity that for every infant that reaches its destination abroad five to ten breeding females have been destroyed. This kind of pressure maintained for a few decades could easily bring the "lazy, hazy, crazy days of summer" to an end for all wild chimpanzees.

Although chimpanzees are generally forest-dwellers, Jane van Lawick-Goodall's remarkable studies were carried out on a population living in a woodland-savanna type of habitat in the Gombe Stream National Park which borders the eastern shores of Lake Tanganyika. Subtle differences are apparent in the behavior of the Gombe Stream chimpanzees compared with those populations studied by Adriaan Kortlandt in East Africa and by Vernon and Frances Reynolds in the Budongo Forest of Uganda that are probably related to differences in habitat. The Gombe Stream chimps are quite often seen easting freshly killed meat—monkeys or young bushpigs; the Reynolds never observed this behavior. Gombe Stream chimpanzees have been observed using "tools" made from grass stems in order to fish for termites and crumpled leaves to act as sponges to soak up water for drinking. This sort of instrumentation has yet to be seen in other wild populations.

These differences between the behavior of forest chimpanzees and woodland-savanna chimps recall Adriaan Kortlandt's theory that chimpanzees have become "dehumanized" (see p. 79), losing their skill in implementation and weaponry and a taste for meat, as a consequence of shifting, with the coming of man, from woodland-savanna and open savanna habitats to the "refuge areas" of the rain forest. Unfortunately for Kortlandt, Jane van Lawick-Goodall's observations do nothing to *prove* his theory; the behavior of the Gombe Stream chimps is just as likely to reflect a progressive adaptation to a *new* environment as the retention of latent tendencies evolved in an old one.

Structurally, chimpanzees are brachiators. They possess long fore-limbs and short hindlimbs with an intermembral index of 107. Their hands are elongated with long fingers and short thumbs, and their wrist joints are specially modified to facilitate brachiation (Lewis, 1969). Their chests are broad, their lumbar columns are short, and their pelvises are wide. On the ground, where they spend the greater proportion of their time, chimpanzees walk in a modified quadrupedal fashion. Even when arboreal, chimps walk quadrupedally along the branches and only occasionally brachiate in the classic fashion. Even so, their long arms and hook-like hands, primarily instruments of brachiation, are not redundant; in their new role as long-distance fruit-pickers, long arms have a decided advantage over short arms. Long arms, however, are less of an advantage on the ground and chimpanzees and gorillas are forced to adopt a compromise locomotor posture somewhere between quadrupedalism and bipedalism—semi-upright but supported by the arms. The adaptations that made the hand of the chimpanzee an effective brachiating device are the very adaptations that prevent its redeployment as a foot; the natural posture for quadrupeds is to place the hand palm downwards on the ground but this chimpanzees cannot do. The chimp and the gorilla, which is faced with precisely the same problem, have reached a compromise solution by supporting the front of the body on the backs of the middle phalanges (or knuckles) of the fingers. This unique feature of the gait has led to the introduction of the term "knuckle-walking" to describe the terrestrial locomotion of African apes. Dr. Russell Tuttle of the University of Chicago proposed the knuckle-walking theory, as it has become known, in 1967. His thesis is based upon the *differences* that he can show between the hand of an orang which is a brachiator but not a knuckle-walker, and the hands of chimps and gorillas which are both. Unlike Washburn, Tuttle (1969) does not support the view that knuckle-walking was antecedent to human bipedalism.

Temperamentally, gorillas are to chimpanzees what Great Danes are to fox terriers. The one dignified, solemn, and reserved, the other noisy, excitable, and extroverted. This personality difference is accompanied by distinctions in physical structure, in habitat, diet, and behavior (Table 10). Adult gorilla males average over three times the weight of adult chimpanzee males; and gorilla females have just over twice the bulk of female chimpanzees. An interesting difference between the two species reveals itself when the average female weight is expressed as a percentage of the average male weight. For chimpanzees this figure works out at 84 percent and for gorillas 58 percent. In other words, there is well marked sexual dimorphism in size in gorillas, while in chimpanzees the degree of difference between the sexes is much as it is in man. Gorillas are solidly built animals with tremendously deep chests and massive, muscular backs. Their hands are much broader, fingers are shorter and the thumb longer than in chimps. A surprising finding, in view of the almost wholly ground-living way of life of the gorilla, is that the intermembral index is higher than in chimpanzees (see *Pan*, Table 8); as a result the gorilla's back, when knuckle-walking, is inclined at a considerably steeper angle than is the chimp's.

Many of the characters that distinguish the head and face of gorillas can be correlated with their great size. The strong bony crests on the skull, the heavy bony architecture of the face, the massive teeth and the protruding jaw are biomechanical responses to large size and a bulky diet (Figure 20). Allometric considerations apart (see Chapter 2), the anatomi-cal differences between chimps and gorillas, such as the shape of the ears and the nose, are quite striking. Chimpanzees have large flapping ears while gorillas' ears are petite and flattened against the side of the head. The nose of the gorilla looks something like a tomato cut in half with thick folds of flesh surrounding the wide open nostrils; the chimp's nose, unlike its exuberant ears, is a much more discrete affair (compare ears and noses in Plates 5 and 6). Western or "lowland" gorillas can be quickly distinguished from the eastern or "highland" race by the presence of a slightly overhanging tip to the nose in the former.

There are two main centers of gorilla population in central Africa, one on the west coast in an area north of the River Congo and east of River Niger; the other in the lowland forest east of Upper Congo (River Lualaba) and in the adjacent mountainous regions of the Mitumba range and the Virunga Volcanoes. The western population is referred to as "lowland" gorillas and the eastern population as "highland" gorillas (although not all of the latter race actually live in the mountainous re-gions; Groves, 1967). The typical habitat of western gorillas is the trop-ical rain forest while eastern gorillas live partly in tropical rain forest, and partly at high altitudes (up to 13,000 feet) in mountain forest (see Chapter 4). Gorillas, unlike chimpanzees, are dedicated vegetarians; no single incident of meat eating has ever been reported in the wild. While chimpanzees are largely dependent on fruit, gorillas favor a diet of coarse vegetable matter (shredded by the large canine teeth) that appears to provide them with the bulk food that their huge frames demand. Also, unlike some chimpanzees, gorillas show little interest in tool-using.

The social life of gorillas lacks the rumbustious elements of chimpanzee societies. The displays of chimps ("carnivals," "rain-dances," and other shenanigans) are noisy, slap-happy orgies of shrieking and screaming, stamping on the ground, thumping tree trunks, and the hurling about of torn-off branches. The gorilla display is more controlled, more dignified, and for this reason, more menacing. Interpreted as a threat or intimidation, the gorilla's display includes a series of ritualized gestures culminating in the classic performance of chest beating. This is executed with the cupped hands beating against the chest, the air-filled lungs acting as a resonating chamber. Chest beating is never seen in chimps who prefer to do their drumming on the buttress roots of forest giants and the hollow trunks of fallen trees. The gorilla's display has been described and analyzed by George Schaller, who spent ten months in intimate contact with the gorillas of the Virunga Volcanoes during 1959-1960. His unique account of their life has been recorded in his book *The Mountain Gorilla* (1963). Schaller's work is being extended at the time of writing by a young Californian girl, Dian Fossey, who has made even greater advances in personal contact with these peaceful animals.

Gorilla groups which vary in size between 5-30 members are relatively stable and cohesive. There is little of the free and easy interchange typical of chimpanzees. Changes in group composition are limited to the birth of infants and the apparently random comings and goings of individual adult males.

Schaller's work has largely dispelled the traditional aura of fear that surrounded the gorilla legend. Gone is the ravening fiend of native gossip and Hollywood B-movies. The monster, which could, and frequently did —according to the 19th century explorer Du Chaillu—mold a hunting rifle with consummate ease into the form of a croquet hoop, was the prototype for King Kong. But times have changed, and the professional ethologist has replaced the impressionable wide-eyed naturalist-explorer of the 19th century; and so instead of the killer-ape on top of the Empire State Building with a fistful of girl, we now have the image of the gentle giant snoozing away the eternal summers of Arcady.

As seems to happen to gentle people everywhere today, gorillas are an endangered species. George Schaller and his colleague, John Emlen, estimated that in the highland gorilla's 35,000 square-mile habitat, only 5000 to 15,000 animals still survive. This represents a population of 0.14-0.42 animal per square mile. Compared with the population figures of 136 per square mile for howler monkeys (which is probably rather on the high side) on Barro Colorado Island, Panama, in 1959, this is desperately low.

Only the orang-utans of Sumatra and Borneo show a lower density. The current estimate of the number of surviving wild orangs is approximately 5000. Approximately 4000 of these are in Borneo and 1000 in Sumatra. The density per square mile is approximately 0.014 for Borneo and 0.006 for Sumatra. Apart from the "empty" areas of the world such as the Sahara Desert, parts of the Arabian Desert, much of the Australian outback, and the northern steppes and tundras, the population density of humans does not fall much below two per square mile even in Central Africa and the Amazonian rain forest.

HOW DO APES DIFFER FROM MONKEYS?

Some of the major differences in structure and behavior between monkeys and apes, other than locomotion, which has already been fully considered in Chapter 5, are discussed below.

The most striking anatomical hallmark of apes is the absence of a *tail*. Like man, apes possess a coccyx but show no external protuberance whatsoever. A not uncommon human abnormality is the appearance of a short tail, but as far as I am aware it has never been reported in an ape.

Monkey tails come in two models—with prehensile tip and without. The fully operational prehensile tail with its area of hairless skin (which is covered in the same sort of lines and furrows as a finger tip and is just as sensitive a tactile organ) is only found in certain members of the family Cebidae of the New World monkeys, the spider monkeys, the howler, the woolly monkey, and the curiously intermediate species the woolly-spider monkey; capuchin monkeys use their tails prehensilely, to some extent, but do not have an area of naked skin. Among Old World monkeys, prehensility is seen in infants of certain species of guenons and mangabeys and is presumably an adaptation to secure a better grip on the mother during the first weeks after birth. The nonprehensile tail, which is the usual model in Old World monkeys and lemurs, is longer than the length of the head and body combined in arboreal forms which use the tail in various ways to aid locomotion. Ground-living monkeys such as baboons, mandrills, and some macaques either have tails which are considerably shorter than the head-and-body length or are altogether absent. The only truly tailless monkeys are *Macaca sylvanus*, the Barbary "ape," and *Cynopithecus niger*, the Celebes black "ape." Two other macaques, *M. arctoides* (stump-tailed macaque) and *M. maurus* (Celebes macaque) have only the merest apology for a tail.

The lack of a tail in apes (and man for that matter) is quite a significant factor and reflects the major postural differences between the apes and the monkeys. Monkeys are quadrupeds and the tail is needed in various degrees to assist locomotion. To be effective as an air-brake or a counter-weight it must be mobile and be capable of a wide range of movements. The muscles controlling these movements are part of the internal pelvic musculature. Apes, fundamentally, are brachiators, that is to say their locomotion is carried out by means of hand-over-hand arm swinging with the body held vertically. A tail would not materially assist this type of locomotion one way or another but, more importantly, the muscles that in monkeys are deployed for tail-movement can be re-deployed in apes for the important function of supporting the abdominal viscera.

In man and apes the "floor" of the pelvis, the surface upon which the whole of the digestive tract and organs and the genito-urinary organs such as the bladder and the uterus rest, is comprised of a muscular diaphragm called the levator ani. The name itself is a misnomer for only a very small part of its total function is to provide support for the rectum and anus. The levator ani comprises several muscular sheets, the most posterior element of which is called the ileo-coccygeus; in quadrupedal monkeys the muscle controls the side-to-side movements of the tail. In the absence of a tail the ileo-coccygeus of apes takes over a new function, that of closing off the pelvic diaphragm posteriorly and thereby materially enhancing its role as a diaphragm to support the abdominal organs. Man's levator ani is very similar to that of apes; its function is equally adaptive to the demands of an upright posture. Whether a primate *suspends* its body from above or *supports* it from below, the effect of gravity on the abdominal viscera is the same and similar adaptations are selected for.

Although there is no fossil evidence one way or the other, it is possible that the earliest human ancestors, after their separation from the earliest pongids, still bore a tail which would only have shortened and finally disappeared as their upright, bipedal posture became firmly established. Could *Ramapithecus* have had a short tail? I think it is quite possible in all seriousness.

The brain of apes is more complex than that of monkeys judged by its size relative to body size and by the arrangement of its components within the cerebral cortex. In terms of behavior, a more complex brain probably implies a more rapid learning ability, a greater retention of things learned, and integration of these memories to produce symbolic or abstract thinking. It is well known that bigger animals have bigger brains. In Figure 16,

Figure 16. Cubes crudely represent the relative brain sizes of man, apes, and monkeys. Correction has been made for differences in body size.

a series of cubes represent the relative brain sizes of monkeys, apes, and man. In this diagram we are making the assumption that monkeys, apes, and man have the *same* body size in order to provide a more significant comparison. We do not know how size and intelligence are related but we believe that a relationship exists. What is likely to be just as important as size, is the density, arrangement, and branching of the nerve cells within the brain, and this is something that as yet we know little about. There is a certain amount of circumstantial evidence that bigness of brain and intelligence go together. It is well known amongst circus trainers, for instance, that the bigger the animal (and therefore, the bigger the absolute size of the brain) the more "trainable" it is. This observation combined with the intelligent exploitation of an animal's natural talent (plus a good dollop of "razzmatazz") goes a long way to explain the "intelligence" of circus animals. It is trainers who supply the real intelligence and animals that supply the appeal by behaving naturally.

Generally speaking, apes are considerably less skilled in the manipulation of objects than monkeys; this is not so much a matter of intel-

ligence as a matter of physical structure. The process of becoming specialized as brachiators cost the apes dearly; it cost them, in fact, their future
as higher primates. Kortlandt's belief that chimpanzees have become "dehumanized" as a result of being driven back to the forests by the evolution
of man-the-hunter, is, I am convinced, far from the truth. Apes lost the
opportunity of acquiring a human-like status by staying in the forests, not
by returning to them.

By remaining arboreal, apes were overtaken by the evolutionary
trend of greater and greater arboreal specialization. This is reflected in
their *hands* by adaptations which modified the very generalized, all-purpose, hand of monkeys into the specialized grappling-hooks characteristic
of the apes. The apes in this and other ways are drop-outs in the race for
primate pre-eminence. Apes do not rate more than a bronze medal in the
hand-efficiency olympiad; monkeys have the "silver" and man, the "gold."
Although monkey hands have a greater potential for manipulation than
ape hands, they suffer from the great disadvantage that they are primarily locomotor structures, inevitably suborned to the needs of weight
bearing. The most revealing index of the hand in terms of manipulative
skill is the opposability index (Table 11). This index expresses the relative proportion of the thumb to the index finger. The higher the index
(that is to say, the longer the thumb or the shorter the index finger) in

Table 11   **Opposability Indices in Some Old World Primates***

Primate	N	Mean	Habitat
Man	31	65	Ground-living
Mandrill	4	58	
Baboon	11	57	
Patas monkey	3	55	
Macaque	20	53	Part arboreal and part ground-living
Mangabey	4	52	
Guenon	13	51	
Gorilla	17	47	
Chimpanzee	8	43	
Gibbon	7	46	Arboreal
Proboscis monkey	5	46	
Langur	50	41	
Orang-utan	23	40	

*Note the close correlation between habitat and the value of the index.

128

Old World primates, the greater is the manipulative potential of the hand.

It is not possible at present to distinguish apes and monkeys simply in terms of the structure of their societies or social behavior. There are two patterns of social structure discernible in ape societies, the family group and the multi-male horde. The general structure of the gibbon or orang-utan society—the family group—is also characteristic of other forest-living, vegetarian primates such as that of many of the lemurs and some of the South American monkeys, the titi monkeys for instance. Gorillas and chimpanzees share their multi-male, nondespotic societies with other forest-living vegetarians such as colobus monkeys, langurs, howlers, and squirrel monkeys. Amongst the monkeys the structure of the social group is extremely variable (Table 4). The one-male unit typical of the hamadryas baboon of Ethiopia is not seen in other baboons of the same genus. Forest baboons run their societies in a different manner from those living in open savannas. Patas monkeys have a social structure that is quite different from that of the closely related arboreal guenons. Amongst the New World monkeys the little information that we possess indicates a similar degree of variability.

Genetic factors quite clearly play a relatively small part in shaping the form of primate societies. It seems that environment is the far more important factor and that, given certain universals of social behavior (which one assumes to be present though they are certainly not yet determined), primates adapt rapidly to new ecological conditions by adjusting their social priorities.

The gibbons, chimps, gorillas, and orangs comprise two closely related families, the Pongidae and Hylobatidae. This taxonomic grouping indicates a close relationship in genetic structure and, hence, in physical and behavioral traits. Yet, in the nature of their societies, apes line up more closely with certain monkey groups than they do with each other.

Apes live longer than monkeys. Reference to Table 1 will show that all *life periods*—gestation, infancy, adolescence—as well as longevity are increased. But more striking than the difference in life periods between apes and monkeys are the differences between apes and man. This observation which appears to contradict T. H. Huxley's famous dictum that the morphological gap that separates apes and man is *less* than that which separates apes and monkeys, will be discussed more fully in Chapter 7.

The size of the infant at birth relative to the mother's weight varies considerably amongst primates. Adolph Schultz (1969) gives the following figures for this relationship. In percentage of body weight of the mother, the infant weighs 5-10 percent in monkeys, 7.5 percent in gibbons, 5.5 percent in humans, 4 percent in orangs and chimpanzees and 2.6 percent in gorillas.[11] The apes have therefore by far the smallest infants and the monkeys the largest. These findings correlate with the fact that monkeys often have serious birth problems owing to the relatively large size of the head compared with the width of the pelvis, but that apes, whose pelvis provides ample room for the small infant, usually have easy and rapid deliveries. Man's problems in this respect are the same as the monkey's. No one

Table 12   Longevity Records of Captive Apes and Monkeys up to 1967*

	Years		Years
Tupaia	5½	*Saimiri*	21
Urogale	7	*Presbytis*	23
Propithecus	7	*Colobus*	24
Loris	7	*Lemur*	28
Perodicticus	8¾	*Mandrillus*	28½
Leontideus	10	*Macaca*	30
Callithrix	12	*Cercopithecus*	31
Tarsius	12	*Hylobates*	31½
Pithecia	13½	*Papio*	32
Galago	14	*Gorilla*	36
Microcebus	15	*Cebus*	38
Ateles	20	*Pongo*	38
Cercocebus	20¾	*Pan*	41

*From Marvin Jones, 1962, 1968.

knows why ape babies should be so small, particularly as their gestation period is, relatively speaking, so long.

    At birth, ape infants are much less mature than infant monkeys and need to be supported by the mothers for the first month of life before they learn to cling to her fur. Monkey infants show the clinging reflex from the moment they are born. The immaturity of apes is reflected in the date of eruption of the milk teeth, the first of which appears in the third month but is often present at birth in monkeys. Delay is also seen in the appearance of the epiphyses of long bones. All the secondary

---

[11] The talapoin monkey *(Ceropithecus talapoin)* holds the record (16.4%) for the largest infant of all primate species according to Dr. Robert Cooper of the San Diego Zoo. Squirrel monkey infants are also relatively large.

centers of ossification are present in the upper limb of the macaque at birth, whereas in the apes only the center for the head of the humerus, the lower end of the radius and for two of the carpal bones have appeared. Man, in both respects, is even less mature; his first milk tooth does not erupt until approximately 6 months, and at birth there are no secondary centers of ossification in the bones of the upper limb.

The longevity of apes and monkeys is not easy to assess; there is no possible way by which estimates can be made in wild populations at present. Long-term field projects such as that initiated by Jane van Lawick-Goodall at the Gombe Stream National Park may eventually provide us with the necessary information about chimpanzees, but we shall have to wait for a long time as apes, in captivity at any rate, may live longer on the average than medieval man. Based on figures collected in 1962 by Sgt. Marvin L. Jones of the United States Army the longevity records of captive monkeys and apes are shown in Table 12.

EVOLUTION OF THE APES

In Chapter 5 we discussed the evolution of the primates in terms of locomotor grades based on major changes in locomotor behavior. Now, specifically, we shall be concerned with the evolutionary history of the apes in the more formal terms of phylogeny. As our object in this book is to trace the roots of mankind we shall confine ourselves principally to the evolution of man's nearest kin, the chimpanzee and the gorilla.

A recently discovered fragmentary, fossilized skull lacking a lower jaw is the starting point for this paleontological safari through time and space. *Aegyptopithecus* (Plate 9) was discovered during the Yale University 1966-1967 expedition to the Fayum, a desert area 60 miles southwest of Cairo. It was found in deposits of Oligocene age thought to be between 26 and 28 million years old. Although its general appearance is more monkey-like than ape-like, its teeth indicate without a shadow of doubt its hominoid affinities. *Aegyptopithecus*, though dentally an ape, was still monkey-like in the size of its brain and form of its skull. A few fragments of limb bones have also been discovered but tell us little about the locomotor behavior of *Aegyptopithecus*; nevertheless I would not hesitate to guess that this primitive ape was quadrupedal but that its hindlimbs were still considerably longer than its forelimbs with an intermembral index of approximately 70-75. Whether or not *Aegyptopithecus* was actually ancestral to the African apes and man, we may never know; but one can certainly be confident that it provides a very plausible ancestral model. *Aegyptopithecus* was in the right place at the right time and had the right dental and cranial structure for such a role.

The general appearance of *Aegyptopithecus* is extremely reminiscent of the smallest of the three species of *Proconsul, P. africanus.* The *Proconsul* group was widespread in East Africa during the early part of the Miocene and the details of the dental and cranial structure are known from several hundred specimens collected from the islands and shores of Lake Victoria. There is little doubt that this group was also widely spread over other parts of tropical Africa but as yet no other deposits have been found. The volcanic activity of the early Miocene of East Africa was particularly propitious for fossilization. Volcanic lava is the ideal preservative for fossils as it provides instant immortality. Seeds, leaves, soft-bodied caterpillars and grasshoppers, and bones alike are preserved by the laval blanket that envelopes them, excluding the air and immortalizing them with a faithful retention of detail, like flies in amber.

Proconsul major, the largest species of the genus, was gorilla-sized as far as can be ascertained from its jaws and teeth and is indeed a possible ancestor for the living gorilla (Pilbeam, 1968). Both *P. nyanzae* and *P. africanus* are likely candidates for the ancestry of the chimpanzee. On the other hand, the possibility that *P. africanus* is a human ancestor cannot be completely discounted. My own studies of *P. africanus* incline me to dismiss it as being ancestral to the hominids. This does not however negate, or even diminish, its importance as a fossil; *P. africanus* provides us with critical information about the sort of bodily structure that Miocene apes and—indeed—early hominids possessed. Undoubtedly the genus *Proconsul* lay close to the common ancestral stem of both apes and man (Figure 17), but it is more likely to have been ancestral to the apes.

Anatomically *P. africanus* was still rather more monkey-like than ape-like. In the form of its skull (Plate 10), the shape and size of its brain, in its inferred locomotion and its hand structure, it finds its nearest living counterpart in some of the more primitive of the langurs of India and Southeast Asia and in the spider monkeys of South America. That *P. africanus* should be such a potpourri of platyrrhine-catarrhine, monkey-ape characters is not really a matter for surprise. In the Miocene, the apes and monkeys had only recently become distinct and it is to be expected that their immediate descendents should show certain characters of both; and that they should both possess characteristics of platyrrhine monkeys is also understandable in the light of the not-so-distant common ancestry

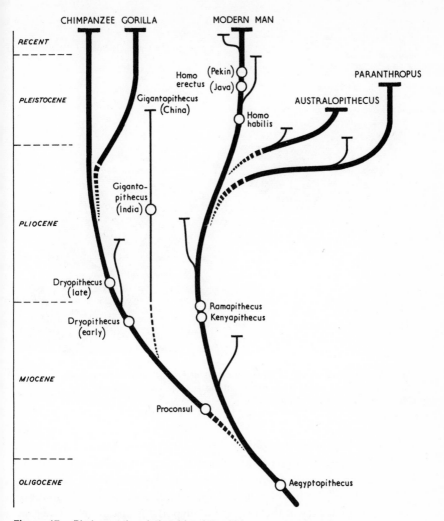

Figure 17. Phylogenetic relationship of the African apes and modern man. Dead end lines represent extinct populations.

of the New and Old World groups.

The dental structure of *P. africanus* is fundamentally of the ape type and differs from it only in the relative smallness of the upper incisor teeth and the fact that the molar cusp pattern is rather more simple and primitive than in living apes.

Ever since Charles Darwin published *On the Origin of Species,* the

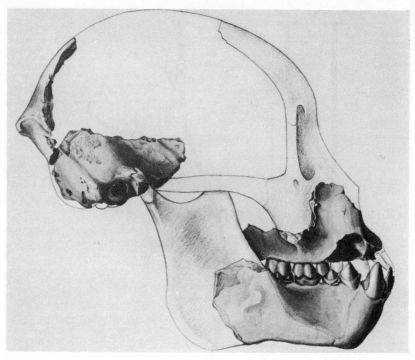

Plate 10. Reconstruction of skull fragments of *Proconsul africanus*. Note that only the clear white areas represent the authors' guesstimates. (Courtesy of *Folia Primatologica.*)

popular view of human origins has been that between apes and man there is a gap in the chain of evolution and if scientists could only find the "missing link" the chain would be complete. From Figure 17 and from what has already been said in earlier chapters, the reader will no doubt be able to see the fallacy of the missing-link hypothesis for himself. The common ancestor of chimpanzees, gorillas, and man was neither an ape nor a man. The relationship between man and the chimpanzee is like that between two brothers, and not like that between father and son. The "missing link" between apes and man of the imaginative writers does not exist. Instead there are literally thousands of intermediate links between man today and his ape-like ancestor of 25 million years ago; and we are aware, perhaps, of one link in a thousand. Furthermore the links are not simply arranged in a linear fashion but as an intricate meshwork of phyletic lines which came together, separated, and came together again. If the link analogy is to be used, it should liken a fossil discovery to a missing link in a suit of chain-mail.

Following the separation of the human-line and the African ape-line, the apes became progressively more and more specialized for an arboreal existence and they gradually assumed the physical characters by which we know them today. Unfortunately we have little direct evidence to tell us how rapidly this process took place. Teeth and fragments of jaws are the only fossil evidence of apes that we possess from this epoch. Thus, the assumption that brachiating adaptations were progressively evolving is largely one of faith; however, there are certain clues in the fossil record that apes were assuming the postcranial adaptations associated with brachiation. In Europe and Kenya during the Miocene, gibbon-like fossils of the genus *Pliopithecus* are well known. They have been identified as gibbons from the characteristics of their teeth. Postcranially they are still monkey-like with intermembral indices in the region of 94. The only living primates with such indices are the fully ground-adapted forms like the baboons, the macaques, the patas monkeys, and the highly arboreal South American spider monkeys. *Pliopithecus* was certainly not ground-adapted in the manner of the living baboons and thus it is to the spider monkey that one looks for illumination. Spider monkeys are the New World analogs of the Old World gibbons in their style of locomotion. They frequently brachiate in gibbon-like fashion assisted by a prehensile tail but, unlike the gibbon, they just as frequently run about quadrupedally. Furthermore it has been recently shown (1965) by Dr. Ankel of Zurich from her studies of the sacrum that *Pliopithecus* possessed a tail, although whether or not it was prehensile is equivocal. We have already suggested in the last chapter that brachiators, evolving from quadrupeds, passed through a transitional stage called semibrachiation and it would appear that *Pliopithecus* fits the bill for this hypothetical stage. Thus the evidence of these Miocene apes strongly suggests that the evolution of brachiating morphology was well underway, but that it had not yet emerged from the transitional or semibrachiating stage. A similar conclusion was reached by my colleague Peter Davis of the University of Surrey and myself when we were studying the forelimb bones of *Proconsul africanus* some years ago. We said then (1959) "*Proconsul africanus* . . . represents a transitional stage in evolution in which an active quadrupedal form is developing the characteristics of a brachiator."

One further small and possibly telling piece of evidence has come to light recently. A lumbar vertebra attributed to the largest of the *Procon-*

sul species, *P. major,* has been reported in *Nature* (1968) by Drs. Alan Walker and Michael Rose. They conclude that this vertebra is very similar to that of modern African apes, the gorilla and chimpanzee. The lumbar vertebrae are important bones in revealing posture and locomotion in primates and, thus, this fragment of evidence may be a pointer to an early Miocene date for the first signs of brachiating adaptations. Several authorities, in fact, regard *Proconsul major* as being an ancestral gorilla on the basis of the anatomy of the teeth and jaws.

It is universally a matter of surprise to laymen that paleontologists can draw far-reaching conclusions on the basis of a single bone. The truth of the matter is that while in some cases it is a perfectly legitimate practice (see footnote, Chapter 9, p. 207), in others it has led to totally erroneous conclusions. A classic example of a fallacious interpretation is the famous tooth discovered in 1922, given the scientific name of *Hesperopithecus haroldcooki* or the "western ape discovered by Harold Cook." Far from being the first evidence for hominoid evolution in North America, as was thought for some time, this misleading molar proved to belong to an extinct peccary, a kind of pig.[12]

A new genus of near-man, however, has been successfully projected on the basis of a single molar tooth. The discovery (and subsequent discoveries of more teeth) was made by Professor von Koenigswald in 1935 in a drugstore in Hong Kong. By Western standards a drugstore is an unlikely place to discover a new primate genus but is perfectly logical to a Chinese for whom drugstores are places where "dragons teeth" can be brought, ground up and taken, three times a day, as a panacea. Since von Koenigswald's original discovery, a number of jaws of *Gigantopithecus* have come to light bearing the identical, gigantic molars that he had used as the basis for his new genus. The latest discovery in 1968 of a huge jaw from the Siwalik Hills of northern India has been described recently in *Scientific American* (Simons and Ettel, 1970).

The next important group of fossils that bear on the evolution of the apes are the dryopithecines (or woodland apes), so called because the original fossil from which the genus *Dryopithecus* and thus the subfamily Dryopithecinae got its name, was found in southwestern France in deposits indicating a warm-temperate but not a tropical climate. The dryopithecines were undoubtedly ancestral apes having the large canines, the

---

[12] This gaffe is much prized by British paleontologists who hoard it as a most effective riposte to be used in the face of accusations from American colleagues of British gullibility over the Piltdown affair!

characteristic long molar tooth row, the deep heavy jaws and protruding faces of living members of the Pongidae (Figure 20).

The dryopithecines were widespread both in time and space. The genus is recognized from the middle Miocene to the lower Pliocene, covering a span of over 5 million years. Dryopithecines have been found in Europe, Africa, the Middle East, India, China, and Russia. Clearly they were a highly successful family living in both temperate and subtropical woodlands. At the present time apes have a distribution that is restricted to the tropical rain forests of Africa, Asia, and Indonesia from 15°N to 5°S of the equator. Clearly this poses a question: whatever happened to the great apes that they are now so geographically circumscribed? What happened, was what happens to ham in a ham sandwich; the competition from man and ground-living monkeys in the grasslands and woodland savannas and from arboreal monkeys in tropical and subtropical forests, simply put the apes out of business. If, in addition to these population pressures, one adds the effects of a changing climate leading to the shrinkage of forest habitats, then it is really not surprising that the ham has been compressed so thin.

Apes have neither the plasticity of man nor the all-purpose, rough, tough characteristics of monkeys when it comes to adapting to a new life zone. They have become overspecialized and their evolutionary "freedom of choice" has vanished forever. The apes today are a tiny relict population of a once widespread and highly successful group of primates amongst which only the gibbon has held its own. Today all the great apes (chimps, gorillas, and orangs) are actually or potentially endangered species. In this chapter I have tried to indicate the extent to which the apes in our attic hold the secrets of our past. It would seem that on rational grounds, if for no other reason, we should make it our policy to see that they survive.

CHAPTER SEVEN

Man in the offing

The first evidence of man in the fossil record that practically all anthropologists would accept without question dates back to the middle Pleistocene epoch some one-half million years ago (Figure 30). Discovered in Java by the Dutchman Eugene Dubois in bits and pieces between 1890-1894, the first man was christened *Pithecanthropus erectus.* Further specimens have been found since and over the years Java Man has come to be accepted as a fully paid-up member of the Club *Homo*—the brotherhood of true man. Nowadays few would quibble with the present designation of Java Man as *Homo erectus.* His most striking character, as his name implies, is that he walked just as upright as modern man does; this much is certain from the evidence of a fossilized femur and from the position of the foramen magnum, the hole in the base of the skull where the vertebral column is attached. The principal reasons why Java Man is not accepted as a member of the inner circle of our own species, *Homo sapiens,* are the size of his brain and the shape of his skull. His skull bones were thick and heavy, his jaw fairly massive, his teeth large (though typically human in form), his forehead short and over his eyes were prominent bony ridges similar, though smaller, to those seen in the skulls of gorillas and chimpanzees. The brain of Java Man, bigger than that of any living ape, falls a long way short of the average for modern man. The largest brain size recorded for *H. erectus* is 1000 cc; this specimen, known as "Chellean Man," was found at Olduvai Gorge in East Africa in 1960.

In 1960 Louis and Mary Leakey unearthed the remains of a species of hominid from Olduvai Gorge, Tanzania, that was subsequently named *Homo habilis.* The specimen consisted of the foot bones, hand bones, the clavicle, and part of the skull and jaw. Since the site where *H. habilis* was found has been dated back to 1.75 million years B.P. (Before Present), it would seem that *Homo* had a considerably longer history than anyone had previously suspected. Naturally this discovery, which entailed a complete re-thinking of the story of human evolution, was bound to come under considerable scrutiny; and indeed it has.

Practically all paleontological discoveries can be described as bones of contention and *Homo habilis* is no exception. This is not the place to air the pros and cons of the *Homo habilis* controversy. Basically, the argument hinges on whether this species is sufficiently distinct from *Australopithecus,* the contemporaneous man-apes of South Africa, and sufficiently advanced towards the human grade to be placed in the genus *Homo.* Dr. Louis Leakey, Prof. Philip Tobias of Johannesburg, and myself, who named the specimen, and a number of other scientists believe that it is so; but there are many that do not (e.g., Robinson, 1965). There the matter rests for the time being. Should *Homo habilis,* as we believe, prove to be a valid designation, then it will have to be accepted as the earliest member of our own genus, one and a quarter million years older than *Homo erectus.* We shall be returning to *Homo habilis* on many occasions in the last three chapters as, like Theseus, we follow the thread of human evolution through the labyrinthine rocks of Miocene, Pliocene, and Lower Pleistocene.

In the meantime, perhaps, it might be as well if we turned our attention to the crucial question, *"What is man"?* Philosophers for centuries have been busily coining neat but futile aphorisms to suit their own predilections rather than to answer the question. Man, said Aristotle, is by nature a political animal. Man, said Rousseau, is a social animal. Man, said Edmund Burke, is a religious animal. Bernard Shaw at least entered the lists with an eye on biology when he observed in one of his plays that "man is simply an amoeba with acquirements." Mark Twain was attempting to provoke rather than instruct when he intervened with his own particular brand of fundamentalism: "I do believe our Heavenly Father invented Man, because he was so bored with the monkey." First prize for prophecy goes to Benjamin Franklin. Man, he said, is a toolmaker. With this profound epigram Benjamin Franklin—quite fortuitously—anticipated by 200 years the definition of man which, because of its cultural overtones, is currently favored.

Let us first dispose of the taxonomic meaning of the word "man." Man is common-name equivalent of *Homo* and modern man is idiomatic for our own species *Homo sapiens*. Mankind embraces all men past and present of whatever species. The ancestral forms of man are given various vernacular names such as ape-men, men-apes, near-men; none of these are really appropriate and should probably not be used. The scientific name for the human family which, of course, includes its early (pre-human) representatives as well as its human ones, is Hominidae. The anglicized version, hominid, is generally used to describe the pre-human stage and can be employed as a noun or an adjective.

The formal classification of Hominidae reads thus:

Family Hominidae Gray, 1825

    Genus *Ramapithecus* Lewis, 1934 (includes *Kenyapithecus* Leakey, 1961)

        Australopithecus Dart, 1925

        Paranthropus Broom, 1938 (includes *Zinjanthropus* Leakey, 1959)

        Homo Linnaeus, 1758

    Species *Homo habilis* Leakey, Tobias, and Napier, 1964

        Homo erectus Dubois, 1892 (includes *Pithecanthropus* Dubois, 1894; and *Sinathropus* Black and Zdansky, 1927)

        Homo sapiens Linnaeus, 1758

No two scientists reading this are likely to agree with the foregoing classification in every respect, but this does not worry them unduly; they are used to the state of affairs where individual scientists use classifications as expressions of their own views. But, agree or not, every scientist will know, by means of this nomenclatorial shorthand exactly which specimens are being referred to. The language of taxonomy and classification is a means of communication and the important thing is to use this language clearly and unambiguously (Simpson, 1963). The *content* of a classification is simply the expression of one man's opinion.

To reach the point where classifications can be made, a list of characters must be drawn up which define a particular species or genus. It is in the selection of these characters that disagreement is rife. The difficulty lies in the fact that evolution is a continuous process in which one form flows into another. There is no hard and fast rule engraved on a Mosaic tablet which states: "Above this line is man, below it is non-man." Each decision of classification has a very large arbitrary element to it. Intermediates, *formes de passage,* cannot be left floating in a nomenclatorial vacuum—they must be attached to one category or another. *Homo habilis,* for instance, has many characters which are characteristic of *Australopithecus* but he also has characters of *Homo.* He cannot be called *"Australopithecus-Homo"* for taxonomic rules do not allow such equivocation, so he must be placed above the line or below the line. Unfortunately the very characteristic that could be adduced most strongly to support the human status of *Homo habilis*—the evidence of a toolmaking culture on his living sites—cannot so be used according to the present rules of the game.

For the anthropologist who is concerned with fossil man, a definition based on anatomy is the only kind that is really of any value to him. The question he has to ask—"How do I know whether the bones I hold in my hand are those of a man or a non-man?"—can only be answered by reference to a diagnostic key which may read something like this:

Homo is a genus of the Hominidae with the following characters: the structure of the pelvic girdle and the hindlimb skeleton is adapted to habitual erect posture and bipedal gait; the forelimb is shorter than the hindlimb; the thumb is well developed and fully opposable and the hand is capable not only of a power grip, but at the least, a simple and usually well-developed precision grip; the cranial capacity is very variable but is, on the average, larger than the range of capacities of the genus *Australopithecus,* although the lower part of the range of capacities in the genus *Homo* overlaps with the upper part of the range of *Australopithecus;* the capacity is (on the average) large relative to body-size and ranges from 600 cc in earlier forms to more than 1600 cc . . . the supra-orbital region of the frontal bone is very variable ranging from a massive and very salient supra-orbital torus to a complete lack of any supra-orbital projection . . . the facial skeleton ranges from moderately prognathous to orthognathous . . . bony chin may be entirely lacking . . . the dental arcade is evenly rounded . . . the first lower premolar is bicuspid . . . The molar teeth are . . . in general small relative to the size of the teeth in *Australopithecus;* . . . the size of last upper molar is generally smaller than and particularly molars and premolars, are not enlarged [from side-to-side] as they are in the genus *Australopithecus.* . . .[13]

[13] This diagnosis of the genus *Homo* is an edited version of that provided in Leakey, Tobias, and Napier, 1964.

For paleoanthropologists, whose concern is the bone and the flesh that once enclosed it, you might think that there would be no problem. But you would be wrong. For a start, no account, like the foregoing, can be written without ambiguities to provide ammunition for sceptics and critics, and so it has proved. But more significantly, these objective criteria are inadequate to tell the whole story. As we have already indicated, the rules of taxonomic description, as they apply to hominids at any rate, are not designed to define a living creature, but rather an inanimate fossil specimen, stripped of its dynamism and its will to live. The real creature is not only wrapped round with living flesh but is participating in the world about it. It is behaving. And in the long run, it is behavior that determines whether or not an animal is "fit" to survive in a Darwinian sense. To meet the increasing awareness of the relevance of behavior, functional parameters are creeping into taxonomic definitions of man, witness the reference to hand function and locomotion in the diagnosis quoted above. But diagnoses are still of necessity based on those aspects of human biology that can be deduced from the study of dry bones. The most relevant characters for a meaningful diagnosis of man do not leave their mark upon the skeleton. The critical watersheds of human evolution, speech, a hunting and gathering economy, the "pair-bond" and its correlate the incest taboo, exogamy and the home-base, are outside the purview of the paleontologist. The archeologist can, however, take over at this point but his potential is limited. Analysis of "living floors," the habitation sites of early man, can be analyzed in terms of the debris much as the picnic site of a modern family that are not particularly beautification oriented, can provide clues to the dietary predilections and even the social behavior of the late occupants. The evidence is circumstantial but provides the basis for inductive reasoning of a limited sort with regard to toolmaking, hunting, fishing, homemaking and so on. In similar fashion the natural debris of the immediate geological horizon can be analyzed by zoologists, botanists, climatologists, and paleoecologists who can come up with a reasonable facsimile of the floral and faunal package and hence the prevailing vegetational patterns of the environment. Finally the human ethologist (or behavior-oriented anthropologist) can illuminate the historical scene by hypotheses derived from primate behavior studies or from human behavior studies; he can work *up* from the primates and *down* from modern man and, given adequate information on the climatic and vegetational milieu, he can posit the probable existence of certain behaviors.

So, you can see that "what is man" is no longer a simple question to be answered by a list of anatomical characters, but is an extremely complex philosophical inquiry that needs the combined talents of anthropologists, psychologists, sociologists, botanists, and the rest to unfold. But first things first.

In the development of an evolutionary progression we have not yet got rid of the apes in our attics. *How do I tell my friends from the apes?* This is the next relevant inquiry. The apes and man belong in one superfamily called the Hominoidea. As has been noted earlier, there are three families within this superfamily: gibbons and siamangs (Hylobatidae), the orangs, chimpanzees, and gorillas (Pongidae) and modern man and all his pre-human ancestors (Hominidae). The other great superfamily of the Old World catarrhines is the Cercopithecoidea, which comprises the catarrhine monkeys (Figure 3). Man's affinity with the apes rather than the monkeys, has been recognized since 1863 when the great British zoologist and staunch supporter of Darwin, Thomas Henry Huxley, underwrote this natural order of affairs with the words, "The structural differences that separate man from the gorilla and the chimpanzee are not so great as those that separate the gorilla from lower apes." Modern science has done nothing to disprove Huxley's, largely intuitive, view of this close relationship except in the matter of life periods. In fact, the modern techniques of chromosome and blood-protein analysis have confirmed it to such an extent that some authorities are even prepared to invite the African apes to join the family of man, if not the exclusive brotherhood of *Homo*.

Nevertheless, close as the relationship undoubtedly is, it is still possible to avoid giving deep offense to one's friends by a failure to recognize their zoological affinities. If you look closely at Figure 18 which shows the human and the ape skeleton, you will observe that, bone for bone, there are few differences that stand out. Differences do exist, but for the most part they are of degree and not of kind; there are differences to be found in the thickness and the length of bones, in the shape of the pelvis and in the relative proportions of the limbs and in the number of vertebrae comprising each functional portion of the spinal column. Apes differ from men in their skeletons just as a truck differs from a sedan. They are both automobiles but, as a result of their different functional roles, they vary considerably in both appearance and performance. Some of the more notable distinctions between man and apes are as follows:

In the sense of the number of hairs per square centimeter of skin, man is as hairy as a gorilla but his hairs are so fine and colorless as to be almost invisible over most of his body. For all the protection they give him he might just as well be naked and it is simply a quibble to deny it. The extent and density of his hairiness (or nakedness) varies between different races. Mongoloids and Negroes have less body hair than Caucasoids. Hairy Ainus (of Caucasoid ancestry) are hairier than most of their Mongoloid neighbors but not so hairy as all that; it is the contrast with their Mongoloid neighbors that makes them seem excessively hirsute. Caucasoids show a considerable range of variation. In certain areas of the body, Caucasoids are hypertrichotic, particularly on the face, the chest, the armpit, and the pubic region. On the top of his head man is hairier (in terms of hairs per square centimeters of skin) than either chimpanzees or orangutans; on his chest his hirsuteness is only slightly less than that of the

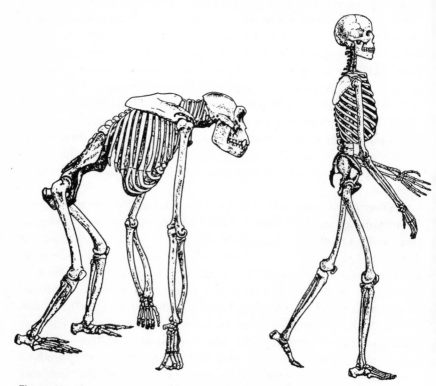

Figure 18. Ground-walking posture of a gorilla (left) and man (right). Note particularly the differences in the shape of the pelvis. (From "The Antiquity of Human Walking" by John Napier. Copyright © 1967 by *Scientific American, Inc.* All rights reserved.)

144

gorilla. The beard is not a highly conspicuous feature in non-human primates except in Diana and De Brazza monkeys, in howler monkeys and in the adult male orang-utan; but even in the latter species it varies between races. The Sumatran orang has the fine, long-haired beard of a Chinese mandarin but the beard of the Bornean orang is unremarkable. Mustaches are particularly favored by marmosets; the mustache to end all mustaches is seen in *Saguinus imperator* whose Emperor Franz Josef adornment is truly imperial. The male patas monkey has the common name of the military monkey at least partly as a consequence of his fierce white "Colonel Blimp" mustache. Of all the apes the gibbon is far and away the most densely furred.

The hairlessness of *Homo sapiens* was claimed by Charles Darwin to be a cooling device, and this still seems to be the most likely explanation. The absence of hair alone, however, is not the key to the adaptive significance of his "naked" skin; hairy coats have an important function in protecting the skin against the injurious effects of strong sunlight.

Man is exceedingly sweaty. His sweat glands, which are essentially of the eccrine variety, cover the whole of his body averaging at about 133 glands per square centimeter of skin; the greatest concentration occurs on the scalp and forehead. Generally speaking it seems that the number of sweat glands are inversely proportional to the thickness of the hairy coat; the less hairy, the more sweaty. Brace and Montagu (1965) have suggested that the ability to reduce body temperature by the evaporation of sweat from all over the skin surface was a critical factor in the selection of hairlessness in early man-the-hunter, whose physical coolness and resilience over long periods of time spent in the pursuit of game in the heat of the day would have been of enormous survival advantage. Furry predators such as lions, leopards, and cheetahs at such times are relaxing in the shade; only mad dogs and Englishmen go out in the midday sun.

The progressive loss of the head hair in human males is a constant source of anxiety to sufferers, but baldness far from being a degenerative disease in modern man is a regular trademark. Certain of the macaques, the uakaris of South America, the orang-utan, and the chimpanzee, are also examples of primates that show progressive baldness associated with maturity. As Doctors Montagna and Harrison have said (1969) baldness is not an affliction but the natural physiological tendency of maturing male primates. Loss of sexual potency is in no way causally related to baldness. Virility is not a commodity encapsulated in hair follicles; if it were mankind would be in a fine fix for baldness starts in the fifth month of fetal life when the hairs of unborn infants start to regress towards the condition seen in the newborn infant.

The regional retention of hair is another matter. As a sexual signal man's body-hair has something to recommend it; it conveys information as to the sexual maturity of the individual, and, because of the differences in its distribution in the two sexes, it conveys information about masculinity or femininity. Its localization on the front of the human body is in accord with current views of the adaptive significance of the face-to-face copulatory position which appears to be biologically ancient in man. As Morris points out, most of the sexual signaling organs, the lips, the breasts, the genitalia, and the major erotogenic zones are sited on the front of the body. The so-called "British missionary position" for sexual intercourse is not only ecclesiastically acceptable but also biologically relevant.

The alternative explanation for retention of body hair in various places is less esoteric and more mechanistic. Hair, by forming run-off channels, aids in the evaporation of sweat from areas of the body such as the pubic region and the armpit where it tends to accumulate in the depths of the skin creases. Man's upright posture which approximates the arms to the sides and the legs to each other, is particularly conducive to the accumulation of sweat in the groin and axilla. There is no need to treat these hypotheses as alternatives. The more adaptive advantages that a character possesses, the more likely it is to be subject to natural selection; most of our bodily structures are multifunctional.

One splendid suggestion to explain the lack of hair in humans was proposed, somewhat speculatively, by the British zoologist, Sir Alister Hardy. As a guest speaker at a sub-aqua club banquet some years ago, he put forward a theory for an aquatic origin of man. The speech was widely reported in the press, and Sir Alister, unwittingly, was credited as the founder of a new theory of human evolution. The "aquatic theory" supposed that man evolved in the region of the seashore or on the borders of large lakes and that his principal diet consisted of seafood—fish, seaweed, and marine mollusks of various sorts. The habit of wading out to sea and groping among the submerged rocks for food would have subjected man not only to selection pressures favoring the upright posture and the development of a thick layer of subcutaneous fat, but would also have favored loss of body hair. It would, at the same time, have favored retention of head hair to keep the sun from burning a hole in his head. Unfortunately for this theory, man like all apes has to *learn* to swim; he does not take naturally to water as several species of monkeys do.

Neither, of course, does the "aquatic theory" explain the retention of hair in certain other parts of the body. Although the "aquatic theory" throws little real light on the evolution of man, Sir Alister's hypothesis serves to focus attention on the importance of the ready availability of water to man's earliest ancestors. No doubt, as Raymond Dart, the discoverer of *Australopithecus*, has always emphasized, a fishing economy could have been an important factor in the life of early man.

In summary, we can say that while man is not strictly naked, he is effectively so, and in this respect he differs from apes. However, it is perhaps more important to understand why he *retained* hair in the areas where it still exists than why he has *lost* it over the rest of his body.

The difference in locomotor pattern between apes, which are brachiators, and man, who is a habitual bipedalist, is reflected in the *proportions* of the limbs and trunk. Apes have long arms and short legs, man has long arms and long legs. Apes have short vertebral columns, man's column is longer. Man's hands are rather short and stubby, apes' hands are long and, in some genera, slender (Figure 19). Apes have short, feeble thumbs, man's thumbs are long, stout, and excessively strong. These are differences of degree but not of kind and they reflect differences in the way of life of apes and man. Such proportional differences are of special significance to hand function and will be referred to again in Chapter 8.

In the *skull* the differences between man and apes are quite notable (Figure 20). For instance, male gorillas have strong, protuberant ridges over the eyes, a crest running fore-and-aft over the top of the skull (saggital crest) and a bony shelf (nuchal crest) at the junction of the back of the skull and its base. Modern man shows none of these characters. Why? Once again we must look for the answer in terms of behavior and environment; in terms, in fact, of the comparative ecology of gorillas and man. Gorillas are vegetarians and have never been observed to eat meat in the wild. Man is omnivorous and eats anything, including meat. The teeth of gorillas are adapted for the heavy chomping that is necessary to obtain nourishment from the rather coarse ground vegetation of celery-like plants that constitute the gorilla's main source of food. As a consequence of this mechanical demand, it is necessary that the jaw muscles, controlling the chomping action of the teeth, should be stout and powerful. In order to give them purchase, jaw muscles need to be attached to bones, therefore the muscles of mastication of gorilla (and man) are, for biomechanical reasons, attached to the skull. The skull encloses the brain and therefore the smaller is the brain, the smaller is the braincase. Gorillas have relatively small brains but they have big teeth and big jaws and, hence, big chewing muscles. In order to provide the necessary area for the attachment of big muscles such as the *temporalis,* the ape skull has extended its available bony surface by developing a saggital crest. The nuchal crest, which meets the saggital crest at the back of the skull like the crosspiece of the letter T, also helps to give attachment to jaw muscles; but its principal function is to provide a bony anchorage for the bulky muscles of the neck (Figure 20). Both crests occur in all adult male goril-

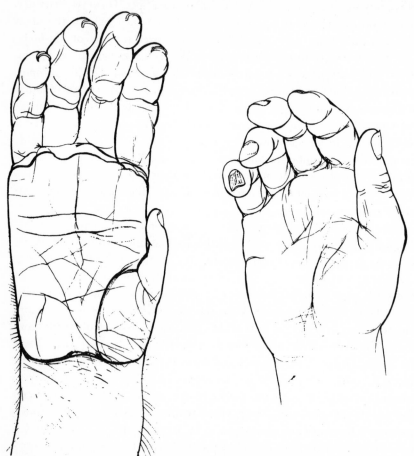

Figure 19.   The hands of chimpanzee (left) and man (right) in a position of relaxation. Note the striking differences in the resting posture, in the proportions of the thumb, and in the length of the fingers. (From *Studies of the Hands of Living Primates,* by J. R. Napier. Courtesy of Zoological Society of London, 1960.)

las and orangs but in only about 16 percent of adult male chimpanzees. Crests are seen in exceptionally large female gorillas but they never reach the massive proportions of the male.

As the brain got bigger during the evolution of man so did the braincase and, thus, the need for special supplementary crests diminished. Furthermore as a result of cooking, man's diet changed in a direction which obviated the need for prolonged chomping of coarse vegetable matter and raw meat. The discovery of fire and cooking led, through the process of natural selection, to reduction in the size of the teeth, and thus to the size of the jaws and jaw muscles. The attachment of the temporalis muscle in modern man has "slipped" from the midline of the skull to the side of the skull (superior temporal line in Figure 20). The human up-

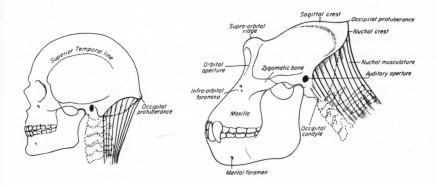

Figure 20.   Contrasts between the human skull (left) and the gorilla skull (right). (From *The Fossil Evidence for Human Evolution*, by W. E. Le Gros Clark. Courtesy of the University of Chicago Press, 1955.)

right posture has also affected the appearance of the skull. The jaws of the gorilla are so large and heavy, and the vertebral column is attached to the skull so far back, that huge and powerful neck muscles are needed to counterbalance the forward and downward drag of the face. As an adaptation to his upright posture, man's vertebral column has shifted its point of attachment to a position well forward on the base of the skull. In this position his head is so nicely balanced that no more than a few gentle touches on the tiller are required from his neck musculature to maintain equilibrium (Figure 20). Thus the large muscles needed in the gorilla to keep the skull hauled back on its condyles, are no longer necessary. As the muscles diminish in size so, too, does the bony flange at the back of the skull, the nuchal crest, to which they are attached. The lightly built and small-jawed skull of man is a relatively recent acquisition. The skulls of *Homo erectus* and the early races of *H. sapiens,* including Neanderthal man, still bear the stigmata of the sub-men, a heavy bony brow ridge below a somewhat flattened forehead. It was not until the arrival of Cro-Magnon man some 30,000 years ago that the human brow assumed its pure and noble form.

In all species of man, including the extinct *Homo habilis* and *Homo erectus,* the canine teeth are small and lie flush with the rest of the tooth row. In the apes, particularly in males, the canines are huge, conical, and projecting. The function of ape canines are doubtfully related to diet. One naturally tends to assume a dietary significance for the shape of teeth but recent studies of the behavior of free-ranging primates have demonstrated that there are other factors acting as agents of dental selection. The value of long sharp canines to male baboons both as deterrents and actual weapons of aggression has a survival value as great as the broadly scooped incisors of fruit-eating apes or the triturating millstones that are the molars of leaf-eating monkeys. Human teeth have their functions in cooperative situations as well as aggressive ones. The smile which reveals the incisor teeth is a come-on signal of considerable force in socio-sexual situations, the "half smile" of sympathy and understanding is a potent grooming device, but the smile that is so wide as to bare the canines is immediately the object of suspicion. "The smiling, damned, villain" in *Hamlet* or the "wolf-like grin" of the classic movie menace are signals of aggression. There are obviously many nuances of the emotive state between these two extremes but these two examples serve to demonstrate the importance of teeth as agents of communication.

In spite of a small canine crown, man oddly enough retains a long canine root. This suggests to many authorities that the long root is an atavism—a left-over from the time that man's canines were as large as those of a gorilla. This is quite possible, but a long root could equally well have been acquired as an adaptation to particular diet. Meat-eating man, lacking the slicing teeth, the carnassials of carnivores, has to do the best he can with what he has got. His rather pathetic cutting armament consists simply of incisors and canines. Visualize how you might cope with, let us say, a glutinous bit of taffy or nougat. You would probably place it at the corner of your mouth and lever it up and down, putting a tremendous strain on the little canine tooth. Early man with his dependence on raw flesh and dried meat must have subjected his canines to a great deal of heavy leverage. For such an activity a long powerful canine root would have been essential.

In many ways the human dentition is poorly adapted for the flesh-eating habits of man, who still retains the tooth form of his fruit- and vegetable-eating ancestors. It is interesting to note, too, that man's digestive system has all the physiological hallmarks of a vegetarian, not a carnivore.

Brain size, as has already been mentioned several times, is partly an expression of body size; but there is an additional factor to be taken into account. As the body size *increases* the proportion that the brain bears to the body *decreases.* For instance a marmoset possesses a brain that is absolutely smaller than a man's but, relative to its total body size, the marmoset brain is bigger than the human brain. It is a measure of the great size of the human brain that although humans are smaller than gorillas their brains are absolutely larger, and of course relatively much larger as well. The extent to which the human brain is larger than the ape brain is shown in Figure 16. We tend to assume that this difference in size represents a difference in intelligence but we have really very few grounds for doing so. One certainty, however, is that there are no structures to be found in the human brain that are not found in the ape's, so the difference is obviously a quantitative one. Charles Kingsley, the Victorian novelist, in his classic *The Water Babies,* caricatured the attempts of Darwin's anatagonists (notably the zoologist Richard Owen) to refute his theories on the descent of man by reference to the uniqueness of the human brain. According to Kingsley, man possessed a "hippopotamus" in *his* brain, something which no mere ape could claim. This rather labored "Punch"-like joke referred to Owen's belief that man, alone, possessed a *hippocampus minor,* a wholly imaginary structure as it turned out. Kingsley's point, though he was a cleric, not a scientist, is well taken. Nevertheless man is more capable than any ape in his ability to learn, to retain the results of past experience and to summon up these memories by instant recall. From this storehouse of knowledge and experience he is able to abstract the essence of a problem and then to express the solution in symbolic terms, representing an abstract idea or quality, by means of speech, written words, signs, gestures, and mathematical formulae. Wherein lies the difference between the brains of ape and man? There is no simple answer to this question. It is not just a matter of size and, as we have already said, neither is it a matter of naked-eye anatomy. Differences, yet to be specifically defined, lie at the molecular level and particularly in the complexity of the branching between the cells of the grey matter of the brain. The more the connections that occur between the sensory cells of the cortex and the association areas, or memory storage areas with which they are closely linked, the greater is the modulating influence that the brain can bring to bear on any incoming sensory experience and, thus, on the motor reaction so elicited. In this way, the variety of possible responses to incoming stimuli (sensory or hormonal) increases the complexity of the behavior of the organism. As an analogy one might imagine that the brain consists of a panel of experts on, let us say, business management. As a consultant body their value increases in proportion to the range and variety of the experience of their individual members and the ability that these individuals have of communicating their knowledge to each other. When faced with a management problem the board *as a whole* can offer a solution that is a synthesis of its collective past experience. The less experienced and integrated the panel, the more inappropriate is likely to be its proffered solution. Experience, integration, and initiation of the appropriate response are the functions of the primate brain but, beyond the assumption that these capacities are somehow related to increased brain size and decreased grey cell density, we are still very much in the dark over the qualitative differences between ape and human brains.

We mentioned earlier that one of the apparent distinctive qualities of man is his use of symbolism. Were the concept of an evolutionary gap between man and the ape to rest on symbolism by means of gesture, then we should be hard put to justify the claim. Jane van Lawick-Goodall, whose work with wild chimpanzees has opened our eyes to the narrowness of the behavioral gap between man and apes in many respects, has shown that symbolism is part of the communication system of wild chimpanzees. Chimpanzees, for instance, will embrace each other, kiss, beg for food, and provide reassurance to strange chimpanzees by hand postures and gestures that are astonishingly reminiscent of the human equivalent. The gestural transmission of an attitude of mind or expression of an internal physiological need in chimpanzees is symbolism just as clearly as are the comfort gestures employed between humans.

Some fascinating studies are in progress on the communicating ability of the chimpanzee by gesture. These studies are being carried out by Dr. and Mrs. Gardner, psychologists of the University of Nevada, with the intelligent cooperation of a young female chimpanzee called Washoe (Gardner and Gardner, 1969). Since Washoe was an infant, all communication between her and her foster parents has been by sign language. A form of the American Sign Language (A.S.L.) is used with modifications made necessary by the limitations of chimpanzee hand movement and as a result of chimpanzee inventiveness. Washoe already has an extensive vocabulary of 34 signs. The signs she uses are perfectly comprehensible although you might say her "pronunciation" leaves something to be desired. Washoe uses combinations of signs such as "listen eat" (at the sound of an alarm clock signaling meal times) and, with the addition of simple pronouns like "I" and "you," something resembling short sentences are beginning to appear. The Gardners hope that Washoe can be brought to the point "where she describes events and situations to an observer who has no other source of information" (1969, p. 671). If and when this occurs, a chimpanzee will have achieved through sign language one of the basic functions of human speech. Already the gestural abilities of chimpanzees, determined from field observation and laboratory study, encourage us in the belief that gesture and sign language preceded verbal communication in man's evolution. Gesture and sign language are in fact still an important part of his communication system and one that he resorts to, thankfully, when this new-found trick of speech lets him down in moments of temporary aphasia.

The difficulties inherent in a comparison of human and non-human intelligence are considerable, mainly for the reason that Washburn, Jay, and Lancaster (1965) pointed out that "learning is part of the adaptive pattern of a species and can be understood only when it is seen as the process of acquiring skills and attitudes that are of evolutionary significance to a species when living in the environment to which it is adapted." A corollary of this is that the more varied the environment, the more a primate must learn if it is to survive, and, conversely, no primate will learn more than is necessary for its survival. Perhaps this explains the placid life of the mountain gorilla. George Schaller, in his study of this species, observed that his gorillas completely ignored inanimate human items, like rucksacks or tin cans that were left casually in their path. Strange objects were outside the gorillas' concept of the necessities of life, so why bother with them? Man, on the other hand has not only an enormously wide range of things and situations that he *must* learn if he is to survive, but through the medium of reading, of night school, of special lecture courses, etc., he voluntarily embroils himself in the task of learning facts and acquiring attitudes that are often quite unrelated to his needs for survival. Who *needs* to attend a course of lectures on "The Lydian Mode and Its Influence on the Baroque of Macedonia"? But plenty of people do! Man possesses the mental abilities to absorb quite unnecessary facts and feel satisfied in so doing that he has been flexing his intellectual muscles. In evolutionary terms, this uniquely human behavior is simply an expression of an adaptive mechanism. Survival of the fittest, as far as man is concerned, is no longer a matter of differential breeding. Breeding can take care of itself. The survival of the human species depends on the intellectual control of the environment exerted by its members. Dobzhansky (1963) has expressed it more elegantly, "Culture is an adaptive instrument which permits the human species to evolve by fitting its environment to its genes more often than by changing the genes to fit its environment." The acquisition of knowledge, both for its own sake and for its conditioning effect upon intellectual muscles, is adaptive for man.

The most significant intellectual advances of modern man that have emancipated him from the forces of natural selection lie in the field of medicine. The insulin treatment of diabetes, for example, and the curative measures now employed in certain diseases of infancy and childhood, have effectively removed a hazard which under the laws of natural selection would have prevented these individuals from reproducing. However, these anti-natural selection advantages are only enjoyed by a small proportion of the total population of mankind. In the context of man's control over his environment it must be remembered that there are a number of equally intellectual "advances" which are destroying life, not saving it. To assume that man has emancipated himself from the forces of natural selection would be quite wrong.

Physiological changes in the adult female sex organs of both man and apes show a 28-day *menstrual cycle*—the end of the cycle being characterized by menstrual bleeding. At approximately the middle of the cycle, a point corresponding with the release of the ripe ovum from the ovary, chimpanzees show a period of sexual swelling during which the external genitalia of the female become grossly enlarged, pink, and protuberant (exciting a mixture of fascination, revulsion, and concern amongst zoo-goers who regularly complain to the authorities that such-and-such an animal shouldn't be on show with such a terrible tumor). Associated with these external genital manifestations of the estrous period of the menstrual cycle, chimpanzees show a heightened receptivity to males and actively solicit intercourse. Neither gorillas nor man show any evidence of sexual swelling or of this particular brand of overt solicitation by females.

In some lower primates the *birth season* is limited to a definite period (or periods) of the year and is closely correlated with environmental factors. Higher primates, however, seem to be capable of breeding at any time during the year, although there is a clear trend towards seasonality (usually associated with the onset of the rainy season) in certain species such as rhesus monkeys. Little is known about breeding seasons of chimpanzees and gorillas in the wild but in captivity there is no evidence of seasonality. For man there is good reason to believe that the environment still has an effect, statistically anyway, on the birth season. According to Zuckerman (1957) the highest birth rate is reached in the northern hemisphere in the first half of the year, and in the southern hemisphere in the second. So far nothing has been said that would help to distinguish one's friends from the apes (the gorilla at any rate) on the score of menstrual cycling or breeding season—but this is because we have not fully considered the phenomenon of receptivity mentioned above. Female chimpanzees are receptive for several days round or about the estrous period; outside these periods there is no sexual encouragement by females nor any inclination on the side of the males. But in our own species the female is, in theory, receptive at all times. In practice, there may well be moments of disinclination but these are probably determined by cultural rather than physiological factors. In man sexual intercourse has ceased to be simply a biological necessity and has in addition become a cultural distraction.

The two principal functional characters that distinguish men from apes concern the human ability to walk upright on two legs and his aptitude for using his hands for activities demanding a high degree of manual dexterity. These are of such significance that they have been given a chapter to themselves. Finally, some of the most important differences between men and apes concern the life periods. Some of these have been already discussed in Chapter 6; here we shall consider the duration of the adolescent period and the life-span.

Just as apes show a slowing of *growth rate* and prolongation of life periods compared to monkeys, so does man compared with apes (Table 1); but in man the degree of difference is even more significant. The most striking differences occur in the prolongation of the juvenile phase and in longevity. The unique lengthening of man's adolescent period can probably be correlated with the cultural advances of modern man; he has more to learn and needs longer to learn it. His whole growth process has, by natural selection, been slowed down to meet this need. We are now reaching the point where education in a technological age demands a longer period of dependency than social and sexual evolution allows for. Today's university student is sexually and intellectually mature, while still technologically adolescent. He is married, restless, and economically deprived. More significantly, he is denied the socio-economic status that his age and sexuality demand. Natural selection *may* eventually operate in the direction of delayed maturity though at present the trend is in the *opposite* direction, towards earlier maturity. One remedy is to meet this biological truth head on and to modernize the system and upgrade a university education so that it marks the start of adult professionalism rather than the end of adolescent schooling. In the progress towards the social evolution of education, student revolts are a grave hindrance; the antagonism they create subverts the exercise of rational thought and reason is replaced by the burning resentment of the establishment whose authority is being challenged by the young males. They act as adult males do in all primate societies when they are so challenged—aggressively. The threat to their position in the hierarchy is one to which the dominant males are not yet ready to submit. The generation gap is a biological phenomenon not a manifestation of a "sick society." There can be no bridging of it, it is an ancient fact of life and is, moreover, of adaptive value for society. Let us hope our educators come to understand this and use the unique talent of *Homo sapiens,* the exercise of reason, to escape from the shackles of his biological past. But understanding must come first or reason simply becomes a euphemism for compromise.

By modern standards the life expectancy of prehistoric man was short; even 40 years would have been considered a ripe old age. During progressive civilizations such as that of ancient Greece, there may have been a small, slow rise in life span. Dr. Lawrence Angel, an anthropologist at the Smithsonian Institution, has discovered that ancient Greeks showed both an increase in body size over their predecessors and an improvement in dental hygiene. In Ancient Rome the *average* length of life was a little over 20 years, in the Provinces a little over 35 years, and in African colonies under Roman rule, 45 years.

It is clear that the average life span of man depends largely on social conditions and the epidemiology of such pandemic diseases as plague and cholera. In medieval England 33-35 years was about the maximum life expectancy; in Europe in the era of the plagues of the 14th and 16th centuries it was 30 years. In the 16th and 17th centuries in the English parish of York, only 10 percent of the common people achieved the age of 40. The aristocracy, on the other hand, were considerably longer lived (Cowgill, 1970). Since then life expectancy has steadily risen. In 18th century England, it was 38-39 years, and in the middle of the 19th century,

40-44 years (males) and 42-48 years (females). In the United States the story is much the same 1830-1860, 41.01 years for males 42.91 years for females; 1893-1897, 44.09 years for males 46.61 years for females; and 1900-1902, 46.07 years for males, 49.92 years for females.

The longevity record for apes in captivity is held by a chimpanzee of the Philadelphia Zoo who died at the age of 41 years. Their potential longevity has been estimated by Riopelle (1963) as 60 years.

This chapter has been concerned with the way that man differs from the apes. I hope that I have succeeded in demonstrating that the interface between the two families, at best, is somewhat blurred. There is no hard and fast rule, no Good Housekeeping seal of approval, which helps to authenticate "man." Man is one expression of the continuity of nature which does not recognize the neat pigeon-holes that we, for our own convenience, like to impose upon the natural world. How do you isolate blue in the spectrum of colors? Whereabouts do you make your cut? Through green? Through indigo?

We classify organisms into species, genera, families, orders, classes, and so on, to express our conviction of the orderliness of nature. But nature is not orderly or precise; it is not concerned with regimentation, only with a continuous, onflowing process of change and speciation, in which the hard lines of demarcation do not exist. Man, as we know him, evolved out of a background of ape-like forms. He evolved in a mosaic fashion, first his posture, secondly his hands, and thirdly his large brain. Did man become Man when he walked upright? When he evolved dexterous hands? Or when his brain was big enough to think in abstract terms and to plan ahead? The decision is quite arbitrary and utterly unimportant. It is better that we think of man as simply a primate emerging from the melting pot of a host of other evolving primates, the possessor of certain abilities which are not *better* in a biological sense than those possessed by any other primate—just *different*. If man wishes to regard himself as special, that's fine. By so doing he is giving expression to a peculiarly human adaptation—that man is the only primate who is concerned about discovering just what sort of primate he is.

CHAPTER EIGHT

Man in the making

The making of man has been going on for some 20 million years or more. Since the ape and human ancestral stocks went their own separate ways, hominids acquired their own set of structural and behavioral characters that, as time went by, distinguished them more and more distinctly from their pongid neighbors.

During the last 30,000 years man's gross anatomy has hardly changed at all. Despite popular belief, man's brain is not likely to get larger and larger and his legs smaller and smaller as a result of thinking too much and walking too little. There is no indication, however, that natural selection is at an end as far as man is concerned—far from it. For instance, natural selection for disease resistance must be a very important present factor in man's evolution. It may be, too, that natural selection has changed direction and is operating on man at some level other than the purely physical. Selection today would be likely to favor certain facets of man's cultural life such as his social relationships and a readjustment of the male and female sexual role. Minor modifications of physical form may conceivably be involved, as in the example discussed below, but radical changes—changes in locomotion, in the number of digits on the hand and foot or the development of a prehensile tail—are highly improbable.

A few years ago, wittingly, I became involved in a minor public controversy on whether or not the human species would eventually become less sexually dimorphic. I considered that such an event was inevitable, but there were others who argued that this view was based simply on a cultural trend (almost subcultural in fact for at that time the long-haired style for men was limited to Britain) and was simply a matter of fashion. I felt it was deeper than that and augured a change in secondary sex characteristics as well as in the behavioral expression of the male-female sex roles.

Sexual dimorphism in size is not very marked in the human species; it corresponds approximately to the degree found in chimpanzees. Human males are between 10–15 percent larger in height and weight than human females. Apart from the primary sex characters, the principal physical differences between males and females do not appear until puberty.

The primary characteristics of sex are the genitalia and reproductive tract, and these differences are present long before birth. Anatomically at an early stage the embryo is neuter; at about three months the cells of the genital swellings differentiate into ovaries or testes and the sex of the embryo becomes overtly expressed for the first time. At birth, the sex is only cryptically recognizable, but as a signaling device to strangers our Western culture has adopted a color code, blue for boys and pink for girls. At an early stage of postnatal development the growth curves for boys and girls are fairly closely matched; after one year of age, girls are generally plumper than boys as they retain their "puppy-fat" longer.

Although principal physical differences await the coming of puberty, differences in personality and temperament, for instance in play behavior, manifest themselves at quite an early age. Rough play and aggressiveness, investigative behavior, and mechanical ability are more pronounced in 4-5 year old males than females. Freedman (1967) attributes this "maleness" to a prepubertal increase in the hormone androgen. Blurton-Jones (1967), as a result of a preliminary ethological study of play behavior in preschool children, attributes the sex differences (which has been observed in infant macaques as well as in humans) to differences of physique rather than to a hormonal effect.

Girls reach the adolescent stage first, an event which is accompanied by a sudden spurt in growth which leaves the boys of comparable age far behind. The male spurt occurs approximately two years later. At puberty, the gonads (testes or ovaries) have completed their development and secondary sex characteristics, which define the adult form, are fully operational. Secondary sex differences include the height, the contours of the body, the relative proportions of the limbs (males have longer forearms relative to the upper arms, for instance), the development of the breasts, the dimensions of the pelvis and the distribution of body hair. For a comprehensive discussion of growth and early development of adolescents, see Tanner's *Growth at Adolescence* (1962).

It would seem that secondary sex characters serve several distinctive functions. The greater height and muscular development, for instance, are related to the traditional male role in society, that of the hunter; and the greater width of the pelvis and the development of the breasts, to the physiological requirements of childbearing. This leaves certain aspects of sexual dimorphism, such as the distribution of body hair, as possible factors of cultural significance, related to sexual attraction and to the strengthening of the pair-bond relationship between males and females.

My argument for future selection operating *against* sexual dimorphism goes like this. Man no longer chooses his mate for reasons of her fecundity.[14] This results from a number of cultural influences, which include the emancipation of women and their altered role in society, the widespread awareness of the dangers of overpopulation, and the changing face of fashion which rejects the Rubensesque and favors the Modiglianian. Man's mate must be something of a chum, slender, and having, preferably, the capacity to earn her own living. Thus, selection on the part of the male is operating *against* the broad-hipped, fleshy, female female.

Sexual selection in females is also undergoing modification. The cave man is through. Females may retain a modicum of the soap-opera-induced urge for muscular rape, but the real issues of sexual attraction are a combination of emotion and pragmatism. "Do I love him enough to live with him for the rest of my life, and can he supply me with the items from Montgomery Ward to which I have become accustomed?" The male male is out, and the weedy but worldly mate is in. The gradual erosion of the male role (the hunter) and the female role (the gatherer and incubator)

---

[14] This sort of generalization is misleading without some qualification of what is meant by "man." It cannot be applied to *all* peoples, as Morris cautioned in *The Naked Ape*, but only to those to whom we refer as "westernized."

must inevitably lead to altered patterns of natural selection in which the intensity of the secondary sex characteristics that provide purely reproductive signals will be diminished. Theoretically, there is of course a limit to the extent to which the narrow-hipped female can survive—a limit imposed by the size of the fetal head. A race of narrow-hipped females could only be biologically viable if the size and maturity of the human baby at birth was reduced. However, in many instances of small or exceptionally narrow-hipped women living and reproducing today, the duration of the gestation period can be artificially modified by induced labor or by Caesarean section. There is no *a priori* reason why, in future human societies, induced labor should not become the accepted routine. Alternatively, natural selection may operate in the direction of shortening the length of the pregnancy by a few weeks, an adaptation which would not impair the "fitness" of the species in view of the present state of medical science in the care of premature infants.

Thus, the vision of diminution of secondary sexual differences and the physical approximation of the sexes is not wholly ridiculous. Man would simply be reverting to a widespread non-human primate characteristic in which size differences (and other secondary sexual characteristics) between the sexes are negligible. Among marmosets and tamarins, for instance, the male and female are identical in size. At least in one respect the sexual roles have been reversed. The male carries the infant on his back at all times except during periods of nursing when the female takes over. In gibbons the size of the canine teeth, which in many non-human primates is a feature that shows extreme sexual dimorphism, is almost identical in males and females. Female spider monkeys in all dimensions are slightly larger than males; and in the majority of arboreal Old World monkeys, for reasons not yet fully understood, the tails of females are relatively longer than the tails of the males.

The current concept of Unisex embodies some of the cultural elements of the trend towards narrowing the sexual dimorphic gap, but its suppositions do not embrace the long-term genetic changes envisaged above. Identity of clothing, hair styles, and leisure-time activities are very superficial expressions of a shrinking sex gap and are little more than a gimmick which will go out of fashion as easily as it came in. Nevertheless Unisex provides a foretaste of a pattern of social change which may eventually become fixed in our culture.

Lionel Tiger in his book *Men in Groups* (1969) looks somewhat gloomily on the prospect of such a social change. The thesis of Tiger's book is that human males form, and have always formed as did their male primate forebears, strong associations or bonds that are of great adaptive significance to man particularly in relation to hunting and war. Women on the other hand do not form comparable bonds and therefore cannot contribute, collectively, to the "important" issues of war and peace. His forebodings of the dire social consequences that would result, for instance, from bond-free women entering the political arena in great numbers, would seem to stem from too stanch an adherence to the idea that the biological universals of the past are immutable and that evolution of man has come to an abrupt halt in the 20th century. The bonding phenomenon is adaptive and is therefore just as open to selection in females as it is in males. With changes in the female social and sexual roles, it seems to me inevitable that bonding, which has been such a successful biological innovation in the male, should, through the process of natural selection, become an equally powerful weapon for sexual equality.

EVOLUTION OF BIPEDALISM

The making of man, a process that is still ongoing, started with the separation of the human-stock-to-be from the ape-stock-to-be. There is no certainty as to when this took place but the consensus favors the late Oligocene or early Miocene (Figure 17). Prior to separation, ancestral man and ancestral apes were part of a common stock, identical creatures (as far as any creatures can be "identical"). They were monkey-like primates with specialized teeth that had already set them apart from their Old World compatriots, the true monkeys. This identity of physical form is likely to have persisted for a very long time so that even a million or so years *after* the separation, the two stocks might have been as hard to tell apart for the ordinary observer as, let us say, *Macaca fascicularis* (the crab-eating macaque) and *Macaca mulatta* (the rhesus macaque). Behaviorally, however, the distinctions might well have been more telling.

The most generally accepted view of the origin of the human stock is that it took place in Africa during the early-to-middle Miocene epoch. The most likely incentive for this important evolutionary step was the changing patterns of climate and vegetation taking place at this time (see Chapter 4), which resulted in the widespread appearance of savannas or grasslands to the detriment of forest cover. This new sort of environment offered fresh and exciting possibilities for a group of hominoids which, no doubt, were suffering from population pressures in the shrinking forests much as certain species of African forest monkeys are today. Life on the ground could have induced the early human ancestors to adopt the bipedal posture, in the first place, as a purely behavioral variation. The advantages of this behavior, however slight initially, would have paved the way for natural selection to favor any genetic modifications that facilitated or improved this method of locomotion by the well-recognized mechanism of genetic assimilation.

There can be little doubt that adaptations for upright walking preceded the evolutionary changes that gave man a dexterous hand. The ability to move about freely and easily and, yet, to have a pair of limbs in excess of locomotor needs, must have been of inestimable value. For monkeys and apes, hands are dual-purpose structures; they function as feet when the animal is moving about and only at rest can they be used as hands. The functional priorities of the foot-hand have resulted in a structural conflict in non-human primates; in man there is no conflict because his hand is almost wholly independent of locomotion.

Common experience will instruct the reader as to the extent in which the arm and hand are used in bipedal walking. Apart from pathological conditions, which require the aid of a stick or a crutch, normal striding involves coincident movements of the upper limbs. At each stride the contralateral arm swings forward to compensate for the twist of the trunk that develops as one or other leg is swung forward (Figure 21–1). When the arms are immobilized by the carriage of burdens this aspect of the gait is impaired with the result that more energy is expended in overcoming the "twist" at the beginning of the next stride (Figure 21–2). The contralateral twist of the pelvis is more marked in women than men. This sexual difference has been exploited in our culture so that the undulating swing of the female hips and buttocks has become a potent visual signal in many civilizations. Man has largely overcome the disadvantage of this impediment by inventing devices that do not involve the hands at all, e.g., rucksacks, papoose slings, and head-balancing techniques (Figure 21–3). However, even though the arms may be involved in bipedal walking, the hands do not support the body weight, so there is no conflict in their anatomical development as primary organs for manipulation.

Man's large brain was a product of the expanding horizons that were opened up by his ability to walk erect and to use his hands. The combination of mind, feet, and hands led him to the use of tools, to the development of hunting, and the establishment of a home base, to the use of

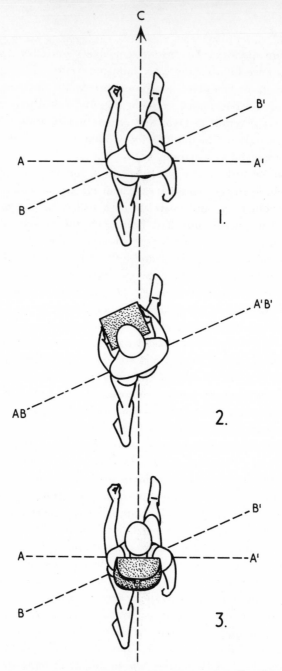

Figure 21. The effect (in exaggerated fashion) of immobolizing the arms on the striding walk of man. A-A' coronal plane of the body.
B-B' the coronal plane of the pelvis.

fire and the discovery of cooking; in fact, to all the cultural trappings that characterize our species today.

We have already discussed the significance of the human brain (Chapter 7), so the remainder of this chapter will deal with the evolution of bipedal walking and the evolution of the human hand—the two principal adaptations that were the making of man.

Human walking is a risky business; without split-second timing man would fall flat on his face; in fact with each step he takes, he teeters on the edge of catastrophe. What one might call the "banana-skins" of life are a constant threat even to the healthy—the rucked-up carpet, the slick of oil, a sudden jolt in the small of the back, are liable to turn even the most dignified of humans into a ludicrous, sprawling object, like a beetle on its back. More serious than these affronts to human dignity are of course the dangers to human health. Under conditions where neuromuscular coordination is impaired—extreme alcoholism, old age, or neurological disease— the hazard is considerably greater. Bipedal walking is risky, not only in terms of the acute accident but also in terms of chronic disease. The "scars of human evolution" as they have been called can be seen daily in their thousands in outpatient departments all over the world —back strains, slipped disks, spondylolisthesis, visceroptosis, the prolapsed uterus, osteoarthritic hips, cartilage tears of the knee, flat feet, and hammer toes. *You* name them—*we* have them!

As for the matter of human birth the whole process is fraught with risk, thanks to the modifications imposed on the human pelvis by the exigencies of upright walking. It is not that man has evolved as a bipedal form too rapidly or that the adaptations to bipedal walking are inadequate, it is simply that dependence on two legs is an inherently unstable proposition.

Amongst the primates, man is not the only one who has adopted a risky way of locomotion. The graceful gibbons, who swing by their arms from branch to branch, have one of the highest accident rates known amongst wild animals. Healed fractures of arms and legs are commonly seen in skeletal collections of these animals. Adolph Schultz (1969) gives a figure of 33 percent healed fractures in a collection of 260 gibbon skeletons. Quadrupedal primates on the other hand, though subject to many hazards of various kinds, are much less liable to locomotor injuries. It seems that the cost of extreme locomotor specialization is a high one, whether one walks on two legs or swings by two arms. Undoubtedly the safest way of getting about is with all four legs firmly planted on the ground; but, though undeniably safer, quadrupedalism has considerable limitations as a way of life.

Man is not the only primate, nor indeed the only mammal, that moves about on two legs. Thus the term bipedalism does scant justice to the uniqueness of the human gait. Birds are bipedal, kangaroos are bipedal, and so occasionally are bears. Many of the giant dinosaurs of the Cretaceous habitually moved about in this fashion; and even among modern reptiles there are still species existing today, like the fringed lizards of Australia, that take to their hindlegs and *run*. Clearly there is no particular problem beyond a purely semantic one in distinguishing between the bipedalism of kangaroos and man; nor is there really any danger of confusing the *occasional* bipedal gait of many primates and the *habitual* bipedal gait of man. When primates such as chimpanzees walk bipedally, the hips and knees are always slightly bent. The arms hang loosely and when seen from behind the walk has a "rolling" quality in which the weight of the body is shifted from side to side as, successively, each foot comes down in contact with the ground. The need to shift the body weight from side to side results from a wide separation of the two feet. When man stands upright his two feet can be brought together so that their inner borders touch each other. Together they form a broad-based pedestal. When man walks, he takes his weight first on one foot, then on the other with a short intermediate phase when both feet are on the ground. At each step he must tilt his pelvis so that his body is perfectly balanced over his weight-bearing leg. The degree of tilt necessary is very small because his feet are sited close to the midline of his body. The feet of chimpanzees and gorillas cannot be brought into contact in the standing position; they are always well separated. When a chimpanzee walks, it is forced to shift its upper body from side to side in a most exaggerated fashion in order to bring its center of gravity over the weight-bearing leg —hence the "rolling" gait. Sailors are traditionally said to have a "rolling" gait. This is because a sailor learns to balance himself on the tilting deck of a ship by standing with his feet well apart to give him a broader pedestal. When a sailor walks the deck in a rough sea he keeps his legs apart and is thus mimicking the walking posture of a chimpanzee, a posture which, through habituation, he often transfers to life on shore. As the instability of drunkenness also demands the same compensatory procedure, a drunken sailor is a sight to see.

This need to tilt the pelvis at each step has led in man to the evolution of a pair of muscles that have no functional counterpart in the non-

166

human primates. These muscles—the gluteus medius and gluteus minimus—are attached above to the pelvis and below to the femur; they are the principal stabilizers of the human hip joint (Figure 22). Although chimpanzees and all other primates possess these muscles their function is not to stabilize the hip, as in man, but to produce extension of the hip in the fore-and-aft plane. The change in the position and the function of this pair of muscles was one of the primary adaptations leading to effective bipedal walking. One consequence of the shift of the gluteus

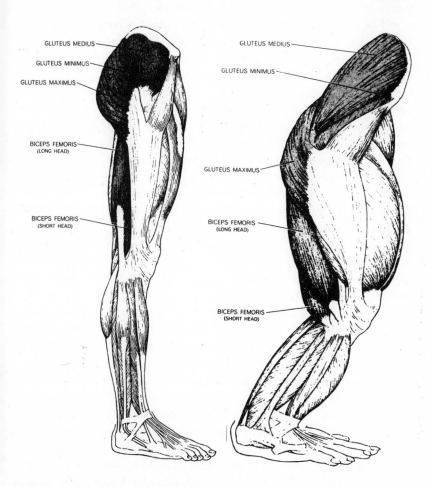

Figure 22. The musculature of the leg in man (left) and gorilla (right). (From "The Antiquity of Human Walking" by John Napier. Copyright © 1967 by *Scientific American, Inc.* All rights reserved.)

medius and minimus to a *new* function was that another muscle had to develop to take over their *old* function (Napier, 1967); this muscle is the gluteus maximus. The gluteus maximus is a large fleshy affair which, in its size and performance, is unique to man. In modern man it does what the gluteus medius and minimus used to do in his ape-like ancestors; it serves to extend his leg in a fore-and-aft direction. This muscle also has an important part to play in extending the upper body relative to the legs; that is to say if one straightens up from a bent position, it is the gluteus maximus that is doing most of the work.

The refinements of human walking express themselves most clearly in the foot, which being at the end of the leg and in contact with the ground should, logically, demonstrate the characteristics of the human walking pattern in the most critical manner. Comparison of the chimpanzee and human foot (frontispiece) reveals a number of differences. Proportionately, chimpanzee toes are much longer than man's but—more significantly—the big toe in chimps is abducted, widely splayed, so that its long axis lies at an angle of 60–70 degrees to the rest of the foot. In man the big toe is adducted—it lies parallel with the other toes. This shift in the position is accompanied by a considerable enlargement of the component bones and muscles controlling its movements.

Although the big toe for some reason is a rather farcical structure, presumably because of its strong associations with gouty 18th century noblemen, it is really one of man's proudest possessions, a true hallmark of his humanity. In its relative bulk and length and in its fully adducted position man's big toe is unique; no other primate has anything exactly like it. The big toe plays a critical role in human walking.

Man's walk has been defined by me (1963, 1967) as a *stride*. This definition provides the means of distinguishing between human bipedalism and the bipedalism of other primates and mammals that occasionally walk upright. Among primates, man alone strides and therefore, one argues, any fossil evidence that correlates with striding behavior, testifies to the human status of the specimen.

Striding may be regarded as the quintessence of human walking, a means of traveling during which the body's energy output at each step is reduced to a minimum by the smooth, gently undulating flow of his forward progress. As such, the striding gait must have been highly adaptive for man-the-hunter who thereby was capable of following wild animals for periods of 12 hours or more with the most economical use of his physiological resources. The fossil evidence of early man from Africa suggests

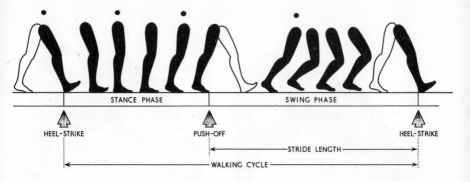

STANCE PHASE			SWING PHASE	

HEEL-STRIKE   PUSH-OFF   HEEL-STRIKE

←————————STRIDE LENGTH————————→

←——————————————WALKING CYCLE——————————————→

Figure 23. Phases of the striding gait of man. Black dots correspond with points of contact on the sole of the foot (white dots labelled 1 to 4 in Figure 24B).

that the evolution of a striding gait occurred well over a million years ago, a period which broadly speaking coincides with man's emergence as a hunter of big game (see Chapter 9).

The various phases of the human walk must be described in order that the functional significance of striding can be understood and that "key" bones, which the paleontologist uses to determine the presence of the function of striding, may be identified.

Human walking is an extremely complicated affair involving many joints and muscles throughout the body and it is beyond the scope of this present chapter to provide a really comprehensive account. Therefore we shall restrict ourselves to studying human striding in terms of the foot, which, after all, is in immediate contact with the ground. Figure 23 indicates the phases of the walking cycle and the relationship of the right leg with the ground during the "swing" and the "stance" phases of the stride.

Figure 24A illustrates the distribution of the static load on the foot during standing; the main areas of weight-bearing being represented by white dots. Virtually half the load on the foot is carried by the heel and half by the ball of the foot. In Figure 24B the sinuous line traces the pathway of load transmission during striding; the white dots numbered 1-4 represent the position of load at four stages of the stance phase of the walking cycle. These four stages are also identified by black dots in Figure 23.

The beginning of the walking cycle, starting from a standing position, involves a potentially catastrophic forward swaying of the body, disaster being averted only by one leg swinging forward at the hip to make

169

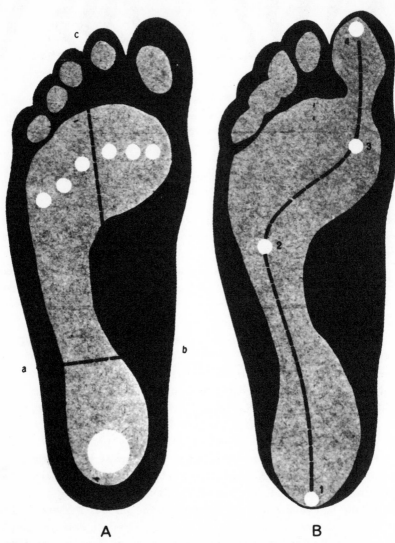

A   B

Figure 24. The sole of the human foot. Light areas represent the parts of
the foot in contact with the ground during (A) standing and (B) walking.
White dots in A represent the principal weight-bearing areas of the foot
when standing. White dots in B represent an arbitrary series of contact points
along the line of weight-bearing (broken line) of the foot during striding; see
Figure 23. (From "The Antiquity of Human Walking" by John Napier. Copyright © 1967 by *Scientific American, Inc.* All rights reserved.)

contact with the ground. The swinging leg strikes the ground, heel first (position 1 in Figure 24B). The contact point is towards the *outer* side of the heel, a fact which the reader can easily confirm by looking at the heel of his own shoe and noting the position of maximum wear. At this point the knee is slightly bent to absorb the shock of the contact. The body continues to move forwards and so also does the contact point of the foot on the ground which is sited along its *outer* border at position 2 in Figure 24B. The body has now moved onward to a position where it is vertically above the foot. But from now on, in the final stages of what is called the "stance" phase of the walking cycle, the foot will fall progressively *behind* the vertical axis of the body. As the body moves over the foot, the point of contact of the foot with the ground also moves, undergoing a remarkably sudden shift in direction to reach position 3 in Figure 24B, a position that is directly in line with the big toe. At this point the *opposite* leg is free of the ground and has started its forward swing. Thus, the whole weight of the body now rests on the stance foot, or rather on the front part of the stance foot. The final element of the stance phase is the "push-off" from the big toe (position 4, Figure 24B). This is the moment of propulsion when the forward motion of the body is supplied by the leg, fully extended at the hip and knee, but still in contact with the ground at the tip of the big toe. By now the *opposite* leg has reached the "heel-strike" position of the stance phase (Figure 23)— and so it goes on.

A complete walking cycle is considered to extend from the "heel-strike" of one leg to the next "heel-strike" of the same leg. Thus a walking cycle is composed of the following events:

Heel-strike right leg   (beginning of stance phase right leg)
Push-off left leg   (beginning of swing phase left leg)
Push-off right leg   (end of stance phase right leg)
Heel-strike left leg   (end of swing phase left leg)
Heel-strike right leg   (beginning of stance phase right leg).

The relative duration of the "swing" phase and the "stance" phase of the walking cycle depends on the cadence or speed of the walk. During normal walking the "stance" phase comprises 60 percent of the cycle and the "swing" phase 40 percent. During normal walking *both* feet are on the ground for 25 percent of the cycle; as speed of walking increases the duration of double weight-bearing diminishes. By the time man is *running* there is a significant period when both feet are off the ground at the same time; and man is airborne.

Not all human walking is *striding*. When man is strolling about casually without purpose or when he is on a slippery surface, he tends to take rather short steps in which both the "heel-strike" and "toe-off" elements are de-emphasized or even absent. The foot is simply lifted off the ground at the end of the "stance" phase and set down flat at the end of the "swing" phase. This is essentially the pattern of walking shown by chimpanzees and other non-human primates when they occasionally resort to bipedalism. But this, for man, is purposeless walking and has little survival value. The adaptive significance of human walking only reveals itself when man is under stress, in competition with nature or with other men. In the early Pleistocene when the better walker was the better hunter was the better provider, striding was the crucial factor for man's survival.

The adoption of a ground-living way of life among the grasslands and savannas of Africa is generally accepted as the event which prodded man's ape-like ancestors up onto their two feet. In recent years, however, an increasing awareness of the impact of environment on primate behavior and a greater understanding of the subtle differences between the various ecosystems has led us to question this traditional assumption. Life in the savannas was no Elysium. Early hominids must have been faced with immense problems. New types of behavior and new sorts of anatomical structure were needed before émigré hominids would be fit enough to meet all the challenges of life in the open. Food would be less easy to come by than in the forest; predators would have abounded and escape would no longer be a matter of fleeing at breakneck speed through the security of treetops high above the ground. If, in addition to these haz-

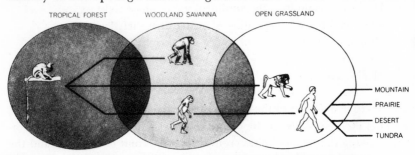

Figure 25. Stylized concept of the ecological pathway followed by monkey, ape, and man through principal tropical biomes in the Old World. (From "The Antiquity of Human Walking," by John Napier. Copyright © 1967 by *Scientific American, Inc.* All rights reserved.)

ards, our early ancestors were in the process of adapting their gait from quadrupedalism to bipedalism, then it is difficult to see how they could have survived the transition. It seems plausible to suggest that man's ancestors did not start living in open grasslands until they were fully bipedal and were able to run and to carry weapons. It would not have been necessary, in the beginning, for them to be able to stride. I believe that this refinement came later and was probably closely linked with the evolution of specialized hunting behavior and the long-distance tracking of wild animals. During the dangerous transition period it is probable that the early hominids were inhabitants of woodland savanna which is the intergrading zone between forest and open grassland (Figure 25). Woodland savanna has enough trees to provide both forest foods and a ready escape from predators; and at the same time the open grassy spaces provide "arenas" where new locomotor skills can be safely practiced and new food-items sampled. In short, woodland savanna would have provided the ideal nursery for evolving hominids, combining the challenge of the open grassland with much of the security of the forest (Napier, 1967).

Viewing the acquisition of bipedalism in terms of evolutionary theory, we must ask ourselves a very significant question. What advantage did this mode of locomotion have for the way of life of early hominids that quadrupedalism did not? We have already suggested the broad scope of the answer. Bipedalism provided a pair of limbs *in excess* of locomotor needs, but it would obviously be of great interest to anthropologists to narrow down the nature of the selective forces of the environment that brought about such a radical shift in locomotor evolution. Before attempting to answer this question it is important to remind ourselves of the nature of man's primate background and consider some of the influences that might be considered as pre-adaptations for bipedalism.

The great paradox of human evolution is that man owes his success as a ground-living form to the fact that his remote ancestors were tree-living primates. Enough has been said in early chapters to make it clear that the basic characters that go to make a primate are almost exclusively derivatives of an arboreal life—the forwardly facing eyes, the short face, the big brain, the posture of the body, the prehensility of the hands, and so on. This view of man's anatomical debt to his remote past has been ratified by many authorities, particularly the late F. Wood Jones and the late William K. Gregory who in the United Kingdom and the United States, respectively, were leading authorities in this field in the period between the two World Wars. *Arboreal Man* by Wood Jones is a book that no serious student of human origins should fail to read.

All primates, with one or two possible exceptions, can *sit* upright, many can *stand* upright without support from their arms, and some can *walk* upright. In other words, we must view the human upright posture not so much as a unique hominid possession but as an expression of an ancient primate evolutionary trend. The fact that one primate species, among 189 others that make up the Order, has taken up bipedalism professionally, so to speak, should not occasion any surprise. The principal trend of primate locomotor evolution, discussed at length in Chapter 5, was from vertical clinging and leaping, through quadrupedalism, to brachiation. The dominant motif of this trend has been an erect body; even among quadrupedal primates the upright *sitting* posture is characteristic. As Wood Jones (1964) said "The human child sits up before it stands; the human stock sat up before it stood." Man has simply exploited the ancient possession of primates by specializing in erectness. The gibbon has also specialized in erectness; he is erect when suspended by his arms, but in so doing the gibbon has lost the potential for further evolution. The gibbon's arms and hands are committed to locomotion. Man, by becoming erect in a terrestrial environment where he supports himself not by means of his arms but by means of his legs, has freed his arms and his hands for other uses. We can now consider just what these "other uses" might have been. What follows is a brief survey of the uses to which living non-human primates put their hands when they adopt bipedalism for brief periods.

Chimpanzees, macaques, and capuchin monkeys have been observed by numerous observers *carrying food objects* in the hands and walking bipedally. Adriaan Kortlandt has seen and filmed chimpanzees carrying off armloads of vegetables raided from native plantations. Jane van Lawick-Goodall has observed and filmed comparable episodes. Japanese workers studying a colony of *Macaca fuscata* on Koshima Island observed how these animals would carry armloads of sweet potatoes (supplied by the observers on the sandy beach) down to the sea to wash them free of grit. Gordon Hewes, an anthropologist from the University of Colorado, was the first to draw attention to the food-carrying behavior of primates. He suggested (1961) that food carriage might have been the principal incentive for the adoption of bipedalism by early man.

Few non-human primates other than chimpanzees have been observed in the action of *throwing objects offensively*. Chimpanzees at

Chester Zoo in the north of England, for instance, have adopted the habit of tearing up clods of earth and hurling them (underarm) at visitors on the other side of the moat separating the chimpanzee "island" from the public. Kortlandt has studied the aggressive behavior of chimpanzees towards leopards and has filmed some dramatic scenes of stick throwing.

Young baboons which lack the large canines of adults have been observed, in play, to use their hands as weapons of defense and attack. A filmed sequence is available where a young pet baboon is defending himself against a couple of dogs in a play situation; it assumes a bipedal posture in order to use its hands to cuff the dogs.

The upright posture facilitates the *carriage of infants* which are too young or too sickly to support themselves by clinging to their mother's fur. In one experimental procedure designed to test the reaction of a squirrel monkey mother to a helpless infant, the arms of the infant were taped to its side so it was prevented from clinging. The mother scooped the infant into her arms and moved away walking bipedally (Rumbaugh, 1965).

Episodes of *tool-using* in chimpanzees and in capuchin monkeys are on record. These are usually carried out in the upright sitting posture, however, so perhaps tool-using is not really relevant as an incentive for bipedalism.

Advantages of bipedalism, other than those resulting from the emancipation of the hands, must also be borne in mind. Among these, far and away the most relevant, is the increased *range of vision* facilitated by the upright posture.

Among African ground-living primates two species, *Erythrocebus patas* (the patas monkey) and *Cercopithecus aethiops* (the savanna monkey), are well known for their upright-standing idiosyncrasy. Both species habitually employ the bipedal posture as a device to augment their range of vision in order to peer over the tall grasses of the savannas in which they live. When bipedal, many species of Old World monkeys use their tails as the "third leg" of a tripod; this behavior has been observed in patas monkeys and in one species of langur, *Presbytis entellus* (Suzanne Ripley, personal communication).

From among all these models, which function is most likely to have provided the principal evolutionary incentive for the uprightness of early hominids? Carriage of food? Carriage and deployment of tools or weapons? Augmentation of visual range? Perhaps it is most logical to assume that it was a combination of *all* these factors which, through natural selection, better fitted early hominids for survival in their new environment. One thing is certain, arboreal life, with its essential climbing component, may through natural selection have facilitated the development of uprightness but it could not have produced the special characteristic of man—his striding gait. Life in the trees, high above the ground, places a premium on the *suspensory* adaptations of primates, whereas the *supportive* adaptations are more likely to have evolved on the ground. The difference between locomotion by bimanual *suspension* and locomotion by bipedal *support* was profoundly significant. There were two directions which the early hominids, with their heritage of truncal uprightness, could have taken in the game of evolutionary "snakes and ladders." One towards brachiation, which they avoided, would have landed them on a snake and sent them spiraling down to extinction; the other towards bipedalism, which they took, brought them to the bottom rung of the ladder that led, step by step, to evolutionary success.

EVOLUTION OF THE HUMAN HAND[15]

Voltaire said of Sir Isaac Newton that with all his science he knew not how his arm moved. Like Newton, we tend to take our hands for granted. Visitors to the Zoo indulge in transports of delight at the way an elephant reaches for an apple with its trunk; they become quite ecstatic at seeing a squirrel or a prairie dog use its paws to eat; but the ineffable capabilities of the human hand are hardly given a moment's thought. Although most of us have been carrying our hands around for several decades, I suspect not many reading this could state—without looking, mind you—whether their index finger is longer then their ring finger or vice versa. Those of you with long index fingers can flatter yourselves that you are among the elite. You possess a progressive human trait. Those of you with short index fingers must take what comfort you can from the thought that tradition is on your side and that *your* pattern (albeit shared with the monkeys and apes) is at least hallowed by time.

The human hand is the chief organ of the fifth sense, touch. With the eye, it is our principal source of contact with the physical environment. The hand has advantages over the eye because it is a motor and sensory organ in one. It can observe the environment by means of touch

---

[15] The substance of this section is adapted from a discourse on the "Evolution of the Human Hand" by J. R. Napier presented to the Royal Institution of Great Britain on 12 November 1965 and subsequently published in their Proceedings, vol. 40, no. 187 (1965).

and having observed it, it can immediately proceed to do something about it. The hand has other great advantages over the eye; it can "see" in the dark. The hands are situated at the ends of long highly flexible limbs which allow the sensory and motor activities to be brought into action at some distance from the body and, thus, the hand can "see" round corners. The hands have one further important function; they are part of our communication system. In the extent to which they are used to communicate—not only words but emotions and ideas—the hands of man are unique in the animal world. Only chimpanzees amongst the non-human primates appear to use their hands to express emotional states. The use of hands in chimpanzees as instruments of communication has been referred to already (Chapter 7).

One hundred and forty-five years ago the Right Honorable, the Reverend Francis Henry, the Earl of Bridgewater, left in his will the sum of £8,000 to the Royal Society. He requested that the money should be used for the purpose of sponsoring a series of monographs, the famous Bridgewater Treatises. One of these entitled *The Hand* by Sir Charles Bell, was a very remarkable study in adaptation of the hand of man and animals, the more so since Bell was writing at a time when, due to the persisting influence of the 18th-century naturalist Buffon, the "imperfections" of animals were a popular device with which to demonstrate the "perfections" of man. The sloth, for instance, was much pitied for "its bungled and faulty composition." Those of you familiar with the sloth will no doubt appreciate that the mistake is understandable. Bell did not fall into this trap; he was clearly aware of the Hunterian principle that "structure was the intimate expression of function" and that function was conditioned by the environment. Lacking the instruction of Darwin (who was at that very time about his business on H.M.S. *Beagle*) Bell was unaware that the catalyst effecting the perfect correlation between structure and function was not a matter of special creation, but rather a matter of natural selection. The full title of Bell's monograph was *The Hand—its Mechanism and Vital Endowments as Evincing Design*. To Bell, the "design" owed more to divine inspiration at a celestial drawing-board than to the trial and error of an earthly workshop. But Sir Charles Bell was a functionalist who appreciated, above all, the relationship between the form of animals and the world in which they lived; in his belief that the mechanism of this relationship was special creation, he was merely reflecting the scientific convictions of his time. It was John Hunter (1728-1793), the great Scottish anatomist and surgeon, who turned our attention from the structure of the hand to its functions. Bell related the function of the hand to the environment and Darwin came full circle by showing us that, through natural selection, the environment gave birth to structure. Thus, the cumulative reasoning of these three men, Hunter, Bell, and Darwin, laid the foundations for the modern, holistic view of the function of the human hand.

Function of the Hand   In order to discuss the function of the hand it is necessary to devise a special terminology. It is meaningless to use the ordinary language of anatomy, for the hand comprises twenty-five joints, totaling fifty-eight possible movements. Some years ago (1956), I introduced a terminology to describe movements of the hands, not in terms of individual joints, but as a living whole.

In spite of the multiplicity of activities of the hand, involving countless objects of various shapes and sizes, its *prehensile actions* can be reduced to two: the *precision grip* and the *power grip* (Plates 11, 12). In the precision grip the object is held between the tips of the fingers and

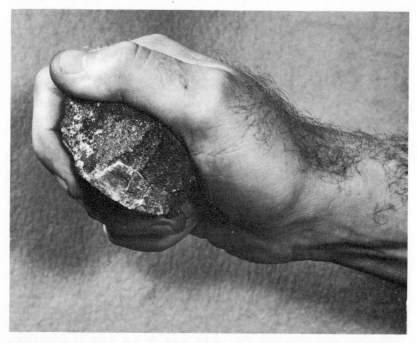

Plate 11.   The power grip, exemplified by grasping an Olduvai pebble chopper. (Photograph by the author. Courtesy of *Discovery.)*

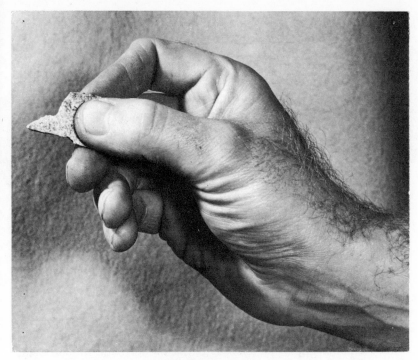

Plate 12.   The precision grip, exemplified by grasping an Aurignacian burin.
(Photograph by the author. Courtesy of *Discovery.)*

the *opposed* thumb; and in the power grip it is held between the under-
surface of the fingers and the palm of the hand, stability being ensured in
the latter grip by the counter-pressure applied by the *non-opposed* thumb.
Which grip is used at any given moment of activity is conditioned by
the nature of the proposed activity. When precision is the more important
or primary element in a given activity, then the precision grip is used; if
power is the primary need, and precision of secondary importance, then the
power grip is employed. For example, a screwdriver can be held in a power
grip position or a precision grip position depending on the size of the
screw and the finicalness of the job. Another example of the relationship
of the nature of the activity and the nature of the grip is illustrated in
Plate 13. This brings us to an important conclusion, that the shape of an
object does not dictate the form of the grip. There is no doubt, however,
that the performance of a given activity, using a tool of any kind, is enhanced when the handle is designed to correspond with the shape of the

Plate 13.   The power grip (left), exemplified by the action of unscrewing the tightly closed lid of a jar, and the precision grip (right), by the final lift off as the lid becomes free. (Photographs by the author. Courtesy of *Journal of Bone and Joint Surgery*.)

hand. Thus, the most successful tools are single-purpose affairs designed *either* for power grip *or* precision grip, but not for both. In most domestic situations a screwdriver is a multipurpose tool, but in professional circles there are power-type screwdrivers and precision-type screwdrivers with handles designed accordingly.

All prehensile movements of the hand can be classed as either power grips or precision grips; there are no others. While this is a useful generalization like so many generalizations, it is not wholly true. There are two other grip patterns both of which are normal to non-human primates but only occasionally crop up in the human repertoire of hand movements. The first is called the "hook grip," used habitually by apes when brachiating. Man employs this grip, which is a variant of the power grip, when strap-hanging on the subway, when opening sash windows or, in the rare instances when it is necessary, for hanging on to the edges of precipices. It is also resorted to carrying a heavy suitcase when the hand begins to tire. The other grip, for want of a better term, is called the "cigarette grip." In man this is self-explanatory but in non-human primates the grip acts as a substitute for the precision grip in certain species. Among New World monkeys for instance, in which the thumb is non-opposable, the "cigarette grip" is commonly employed to pick up small objects between the second and third digits. Old World monkeys, which possess an opposable thumb, use this grip less often, but apes such as the orang-utan, whose thumb is so short as to be almost useless in opposition, very often grasp small objects in this manner (Napier, 1960).

The essential component of both power and precision grips is the thumb. The mass of muscle at the base of the thumb, known as the mount of Venus or more prosaically the ball of the thumb, is composed of a series of small muscles that acting together produce a rotatory movement by which the thumb swings inwards toward the palm. This movement is known as *opposability*. For the movement to be of functional significance the thumb must oppose something. In man, the most precise function that the hand is capable of is to place the tip of the thumb in *opposition* to the tip of the index finger so that the pulps of the two digits make maximum contact (Plate 14). In this position, small objects can be manipulated with an unlimited potential for fine pressure adjustments or minute directional corrections. Opposition, to this degree of precision, is a hallmark of mankind. No nonhuman primate can replicate it. Although most people are unaware of the evolutionary significance of this finger-thumb opposition they cannot be unaware of its implications in international sign language; it is the universal gesture of human success.

Plate 14.   Opposition of the thumb and index finger. This posture is unique to the genus *Homo*.

Evolution of Precision and Power Grips During primate evolution, the power grip was the first to appear and the precision grip came later. This phylogenetic succession can be observed during the development of the human infant whose precision grip only becomes effective long after its power grip has been established.

Rather surprisingly, in the number of digits, in the relative lengths of these digits, and in the number of bones comprising it, the human hand is almost identical with the hand of *Notharctus*, a fossil primate of the Eocene, which lived some 50 million years ago (Figure 1). This suggests that during the evolution of the hand, changes of function have been rather more significant than changes of structure.

A comparison of the hands of living prosimians, New World monkeys, Old World monkeys, and man provides us with a gradation of hand function, from the simple to the complex, which broadly corresponds to what is known from fossil evidence of the successive stages of the evolution of the human hand (Plate 15).

The dual concept of hand function (power and precision) does not really operate at the *prosimian grade*. Lorises and lemurs show a single prehensile pattern whether the hand is being used to grasp a support or to grip an item of food. Prosimian hands function on the mechanical grab principle. Open—close, and when the grab comes up whatever is inside is a bonus. There is relatively little selective use of the hand in relation to a specific function. The prosimian thumb operates on a hinge mechanism but, being set at an angle to the hand, it opposes the fingers as the jaws of a pair of pliers "oppose" each other. As we have seen the true opposability of the human thumb involves a large element of rotation which is totally absent in prosimians. The "opposition" of the prosimian thumb is called "pseudo-opposition." When grasping a vertical support, prosimians show the characteristics of an incipient power grip.

The *platyrrhine monkeys* show relatively little advance on the prosimians as far as duality of hand function is concerned. Anatomically platyrrhine monkeys do not have opposable thumbs. In fact there is an even less advanced degree of pseudo-opposition than in the prosimian thumb. The thumb of prosimians is well separated from the rest of the digits but the New World monkey's thumb and fingers are lined up together in parallel. In spite of this apparently retrograde step the platyrrhine hand shows considerable functional advances over the prosimian hand, even though precision and power grip are barely differentiated.

The axis of their "power grip" lies transversely across the palm and only very occasionally are platyrrhines seen to grip the branch with the thumb on one side and fingers on the other in the true power grip posture. Anatomically their "precision grip" is not a precision grip as it has been described for man, as the thumb is hardly ever involved. When handling small objects some platyrrhine monkeys such as capuchins, uakaris, spider monkeys, and woolly monkeys use the adjacent sides of the thumb and index finger or the index and middle fingers, just as one might pick up an object with a pair of blunt scissors, using them in the manner of forceps. Alison Jolly (1963), who has carried out an extensive study of the hands of prosimians and New World monkeys, considers that the chief advance in the manual ability of platyrrhines rests on the increased use and sensitivity of the finger tips.

Among *catarrhines* the power and precision grips are fully differentiated. Arboreal catarrhines, such as langurs, guenons, and mangabeys, frequently, though not always, use the power grip when walking along branches of trees with the thumb on one side and the fingers on the other. Ground-living catarrhines are less inclined to use a power grip in trees preferring to walk with the whole hand on top of the branch; however, they consistently use a power grip when grasping large objects.

The anatomy of the catarrhine thumb is, in almost every respect, similar to man's. The thumb is opposable and the index finger possesses a considerable degree of independent movement. It is undeniable, however, that monkeys do not use their hands as efficiently as man. As I have already stressed, this is principally because the hands are still, essentially, feet and the range of activities to which monkeys put their hands are severely limited by the functional requirements of locomotion. The environment does not demand of a monkey the high degree of manipulative skill that it demands of man. Monkeys' hands are not subject to the same selection pressures that have operated on the human hand. These selection pressures have induced in man a very high degree of sensory discrimination and nerve-muscular coordination, with the result that the central nervous system can call upon the hand to execute manual acts of a delicacy and precision quite impossible for monkeys.

While the Old World monkey thumb is certainly opposable, opposition between it and the index finger is not nearly as functionally effective as it is in man. In order to approximate the pulp surfaces of the thumb and index finger in the posture shown in Plate 15, a posture that is unique to man, it is necessary that the lengths of the thumb and index finger should be compatible in their proportions. This proportion can be

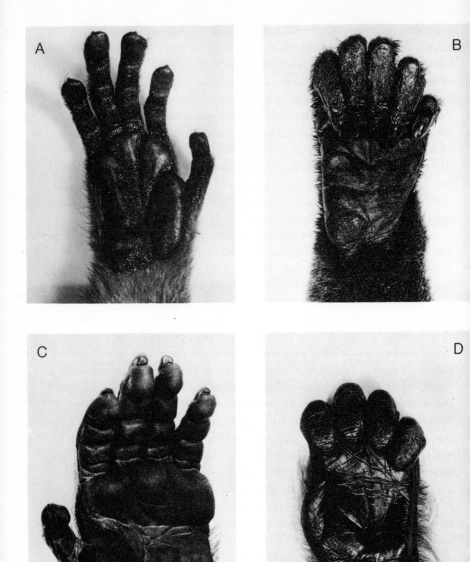

Plate 15. Hands of living primates. (A) Lemur. (B) Woolly monkey.
(C) Macaque. (D) Orang-utan.

184

quantified by means of an opposability index which expresses the length of the thumb in percentage of the length of the index finger (Table 11). The opposability index of man is 65, langurs—41, guenons—51, macaques—53, and baboons—57. It will be seen that ground-living monkeys have higher opposability indices than the arboreal forms. From this index one might make the assumption that ground-living selects for a greater equality in the length of the index finger and thumb. The baboon, which is the most ground-adapted cercopithecoid, and man, who is the most ground-adapted hominoid, have undergone parallel evolution in this respect.

So far, no mention has been made of the ape hand. The reason is simply that the ape hand has no bearing on the evolution of the human hand. The extreme specializations of the hands of the apes, such as the gorillas and chimpanzees, as far as we know took place *after* the separation of the hominid and pongid stocks. Among the apes, whose locomotor specialization is the style known as brachiation, there developed an adaptive conflict between the needs of locomotion and manipulation. Locomotion won. Locomotion by means of arm suspension demands an elongated hand with elongated digits bent into the form of a hook. A long thumb in such circumstances is an embarrassment and, consequently, the same selective pressures that were operating on lengthening the fingers were also operating to keep the thumb short. Tuttle (1967) provides some revealing figures for the muscular proportions of the ape and human hand. For instance, he estimates that the muscles controlling the thumb in apes constitute 24 percent of the total musculature of the hand; in man the proportion is 39 percent.

With the evolution of the precision hand in man, an event which occurred sometime during the Pleistocene, we have reached the end of the road as far as the evolution of the human hand is concerned. Subsequent advances in performance were largely technological. Techniques improved with the advent of special tools, of special tools designed to make special tools, and finally, in this era of automation, of special tools to create tools from which other tools can be created automatically. The day of the craftsman who uses his hands to create his own tools is long past. Now a craftsman's tools are made by industry—to a set and standard pattern—and the result, particularly in the domestic market where consumer demand is imprecise and uncritical, is frequently monstrous. Shapes of handles are dictated by modern concepts of design and "packaging" and have little or no relationship to the shape and function of the hands that are supposed to use them.

For all its achievements in science, art, and industry the hand of man is, to quote Charles Darwin, part and parcel of the indelible stamp that "man with all his noble qualities and god-like intellect still bears to remind him of his mammalian origins."

CHAPTER NINE

Man alive

The fossil record at present offers several hints, but no clear-cut solution, to the time and place of the separation of the hominids from their pongid ancestors. Generally speaking, anthropologists fall into two schools. The *early school* is in favor of evolution of man some 20 million years ago, and the *late school* claims that it occurred some 5 million years ago, more or less.

The late school, fashionable in the twenties and thirties among the adherents of Sir Arthur Keith's "brachiating theory" of human origins, has few supporters today though their number includes several distinguished anthropologists. Some modern scholars of the "late" school like Dr. S. L. Washburn of the University of California at Berkeley have retained the "brachiating theory" and superimposed upon it Tuttle's "knuckle-walking theory" which holds that man evolved from an ape-like ancestor that walked on its knuckles as chimps and gorillas do today. The difference between the two theories is largely a matter of which ape you wish to be descended from. Keith (though in his later years he came to reject his earlier views) saw man as being descended from an ape-like, arm-swinging, wholly arboreal gibbon. Washburn sees man as the de-

187

scendent of an ape like the chimpanzee or gorilla, which spends much of the time on the ground, walking on the backs of its knuckles. Both sorts of apes have the same kind of anatomy as far as general bodily structure is concerned—the outcome of a specialized life in the trees in which arm-swinging (or brachiation) constituted an appreciable part of their repertoire. Washburn's stand in this matter was quite clearly stated in his Huxley Memorial Lecture given in London in 1967: "It appears that our ancestors were arboreal apes for many millions of years, that they then shared a common knuckle-walking stage with the ancestors of the chimpanzee and gorilla, and that only later they became bipeds."

The fossil evidence for the brachiating origin of man is non-existent, but recent studies of the immunological reactions of human and ape albumins by Vincent Sarich, also of the Berkeley campus of the University of California, suggest that the human stock separated from the ape stock no more than a mere five million years ago. Although no one questions the zoological closeness of apes and man as determined by their anatomy and as confirmed by the similarities in chromosome pattern, hemoglobin pattern, and serum protein pattern, the criteria by which the degree of closeness can be quantified in terms of the length of time that has elapsed since the two stocks separated are still highly speculative.

The early school has a certain amount of fossil evidence on its side though this, it must be admitted, is still somewhat thin on the ground. Furthermore, the early school is still undecided upon the nature of the model from which man's remote, direct ancestors could have been derived. Thus there are those like Washburn who would derive man from a fully-fledged brachiator; there are others, like the author, who would prefer that his ancestors were semibrachiators. There are some like W. L. Straus, Jr., of the Johns Hopkins University who would favor a monkey-like quadrupedal heritage; and finally, there are those (though probably few, today), like the late Professor F. Wood Jones, who would go the whole hog and derive man directly from a vertical clinger and leaper like the tarsier. The pros and cons of these hypotheses are too protracted to discuss in detail here so, claiming author's privilege, let me suggest an interim solution that I believe best accords with the present state of our knowledge.

Man is closely allied to the apes and on that there is no disagreement. Thus, any theories claiming that man evolved directly from a true monkey or a tarsier are quite unacceptable. The issue is simply this: did man evolve from a fully fledged brachiating ape or did he evolve from the remote ancestors of modern apes which were neither fully fledged nor brachiating? Most scientists feel that the "brachiating" theory of origin is no longer acceptable because it would involve too many evolutionary reversals to convert an ape-type anatomy into a human type. There is in fact no *a priori* reason why structures once evolved cannot de-evolve. Selection can operate against structures as well as for them; but the sheer magnitude of the reversals necessary to convert an ape into a man makes the theory implausible. Neither is the fossil record in favor of such a dramatic series of retrograde steps.

The alternative theory that man evolved from a pre-brachiating ancestry can now be considered in anatomical terms. Working from top to bottom, the form of the human brain is more ape-like than monkey-like, so is the balance of the skull on the vertebral column. Man's teeth show the primitive characters of early apes and not the specializations of modern apes. Man's vertebral column, his thorax, the musculature of the chest and abdominal cavity and his pelvis are more ape-like than monkey-like. The proportions of the limbs, on the other hand, are more monkey-like than ape-like, so are his hands and so are his feet. Biochemically, man is allied to the apes rather than to the monkeys. What do all these apparently contradictory pieces of evidence add up to? Simply this, that the ancestors of man are not to be found amongst modern apes but amongst species of apes that were still at an evolutionary stage when structurally they were more monkey-like than ape-like. The ancestors of man were "dental apes" that still retained a gait (and therefore the body form) that was closer to that of modern quadrupedal monkeys than to that of modern brachiating apes. Subsequently, in spite of a common heritage, ape ancestors and human ancestors went their different ways. Pongids evolved into arm-swingers and hominids into bipeds. It should be noted that *both* these ways of life involve the upright posture. Apes are upright when suspended by their arms and man is upright when supported by his legs. Herein lies the explanation of the many similarities between apes and man in the *axial* skeletons (vertebral column, thorax, and pelvis) and the many dissimilarities in the appendicular skeleton which comprises the arms and the legs.

EVOLUTION OF THE HOMINIDS

From the Oligocene epoch of Egypt, as we have noted already, there is the fossil evidence of (1) possible emerging monkey stock (*Parapithecus*),

(2) an emerging gibbon stock (*Aeolopithecus*) and, (3) an emerging great ape stock (*Aegyptopithecus*). Elwyn Simons has expressed the view, with very proper caution, that there is evidence of a possible forerunner of man in the assemblage of fossils that this fruitful area has yielded up. This species, called *Propliopithecus haeckeli*, was discovered early in the 20th century by a fossil hunter named Richard Markgraf.

A well-known series of fossils known as *Pliopithecus,* and thought to be ancestral gibbons, had already been discovered in early Pliocene fossil deposits in France. *Propliopithecus* was believed to be ancestral to this group—hence its name. Simons, however, has demonstrated that the teeth of *Propliopithecus* are unlike those of gibbons and more evocative of the human pattern. This opinion, though bold and interesting, is highly speculative as Simons is the first to admit.

The early Miocene in East Africa was, apparently, the heyday of the evolving apes. One genus named *Proconsul* (after a well-known vaudeville performer called Consul, a chimpanzee who used to ride a bicycle, smoking a fat cigar and dressed in immaculate white tie and tails), was particularly well represented. Three species of this genus are known. The smallest species, *Proconsul africanus,* discussed in detail in Chapter 6, was possibly antecedent to both the chimpanzee and the gorilla, but the possibility that this species provided the common stock from which man and modern apes emerged, cannot be wholly discounted at this stage of our knowledge. My own view already expressed in an earlier chapter is that *P. africanus* is not antecedent to man but lies close to the point of separation of the pongid and hominid stocks (Figure 17).

The earliest specimens that seem to have a direct bearing on human evolution are classified as two species which are at present placed in separate genera: *Kenyapithecus wickeri* discovered by Louis Leakey at Fort Ternan in Kenya, and *Ramapithecus punjabicus* discovered in the Siwalik Hills deposits in northwest India in the 1930s by G. E. Lewis of Yale University. Dr. Elwyn Simons has suggested that the anatomical differences between *Kenyapithecus* and *Ramapithecus* do not justify their separation into two genera. If they are united, the genus name will be *Ramapithecus* because in zoological nomenclature the older name retains its priority. For the moment it is convenient to retain both names if only to remind us that these early hominids were apparently geographically widespread.

Study of the zoogeographical map of the world during the Tertiary (Figure 12) indicates that at various times during the Miocene a land

Plate 16. "Jimmy," a young adult orang-utan in bipedal posture a Chester Zoo, England. (Photograph by Doris Sorby.)

bridge connection existed between East Africa and northwest India, crossing the site of the present Red Sea, the Arabian land mass, and the Persian Gulf. During the Miocene *Kenyapithecus* and *Ramapithecus* are found, respectively, at the African and Indian ends of this bridge, *Kenyapithecus* being of a slightly earlier date. Hypothetically, one must envisage a progressive population of early hominids (*Kenyapithecus*) moving away from a woodland-savanna habitat, gradually increasing their territorial range until they had entered the African end of the corridor. After many thousands of years of eastward spread, they debouched into northwestern India. At the end of the Miocene, the landbridge connection was broken as the waters of the Red Sea and the Gulf of Aden joined together to isolate effectively Africa from Eurasia. At this juncture, the once continuous populations of *Kenyapithecus*-like hominids would have become rudely chopped into two populations, similar but separate. After several thousand years of geographical isolation, the two populations might have grown sufficiently distinct to be classed as members of different species. This hypothesis recalls the model discussed earlier (Figure 2) whereby separate populations of the same geographically isolated species undergo genetic change evolving progressively into subspecies, species, genera, and even higher taxa. The Indian form and the African form are so similar that it is very likely they will ultimately be regarded as being two species within the same genus and not in different genera as they stand at present.

The conclusion that *Kenyapithecus* and *Ramapithecus* are closely related to the human stock derives from the following facts: (1) the teeth are arranged in a parabolic curve as in man and not in a rectangular form as in apes;   (2) the incisors and canines are small as in modern man; in the apes the canine is long and projecting and the incisors are broad and spatulate;   (3) the facial skeleton is short and rather straight unlike the more projecting muzzle of the apes (Figure 20). The characters of the teeth and the shortness of the face has suggested to some authorities that *Ramapithecus* was an upright bipedal form that was able to use its hands for feeding purposes and was, therefore, possibly a carnivore and tool-user. This deduction from such sketchy evidence (none of which relates to the body below the head) is of course highly conjectural. As we have already seen, it is unsafe to prophesy the shape of the body from the form of the teeth. Early Miocene apes, for instance, had the teeth of apes but their bodies were more monkey-like than ape-like, a state of affairs that could not possibly have been foreseen from the appearance of the teeth alone.

If *Ramapithecus* and *Kenyapithecus* are ancestral hominids, which present evidence distinctly favors, then they are also ancestral to the Australopithecines of South Africa living some 10 million years later.

The Australopithecines are a well-known assemblage of South African fossils. The original discovery that led to the australopithecine bonanza was made in 1924 by Raymond Dart, then a young Professor of Anatomy of Witwatersrand University in Johannesburg. The discovery at Taung in Bechuanaland of a skull of a young individual, subsequently given the scientific name of *Australopithecus africanus,* was of immense importance. The scientific community, however, was not overimpressed by Dart's discovery. Largely, as one now sees in retrospect, this was the result of inadequate communication between scientists half a world apart (see Professor Le Gros Clark's account of Dart's discovery in *Man-Apes or Ape-Men* published in 1967); but also partly because the Piltdown skull, at that time widely accepted by scientists, seemed to indicate an altogether different sort of evolutionary history for man. Moreover, of course, the Piltdown skull came from Sussex, England, an infinitely more respectable place for man to have evolved than the African colonies.

The Taung specimen was labeled as a fossil ape by most scientists and found no place in their theories of human origins. It was only after the second World War that Robert Broom of the Transvaal Museum, Pretoria, who had been collecting fossil material from several sites in the Transvaal, published his discoveries in a monograph, that scientists began to wake up to the true importance of the Australopithecines. In his book, Professor Le Gros Clark describes his own conversion after studying the material in South Africa in 1947 and the subsequent acceptance of australopithecines by the scientific community. In spite of seemingly unequivocal evidence of the near-human nature of this group, however, there are still in Britain and elsewhere no doubt (South Africa for instance) some who remain steadfastly opposed to this view.

Broom's assistant, and later his successor at the Transvaal Museum, was Dr. J. T. Robinson who has made considerable contributions both to the treasure chest of South African pre-human fossil collections and to their taxonomic interpretation. At the present time we have a large collection of teeth, jaw and skull fragments, several fairly complete skulls, four partly complete hip bones, bones of the vertebral column, bones of the arm and leg, a few hand bones, and a few foot bones. Study of these fossils has produced the following list of characters for *Australopithecus:*

(1) Small brain case, large jaws and teeth. Brain volume estimated at 375-650 cc. The modern gorilla range is 350-750 cc with an occasional individual exceeding this. The range for the genus *Homo* (*Homo erectus*) is 775-1225 cc. *Homo sapiens* has an extraordinarily wide range extending from approximately 1000 cc to 2000 cc but averaging 1300 cc (Figure 26). While australopithecine skulls indicate a smaller brain capacity than large gorilla skulls, it must be remembered that brain size and body size are closely related. Gorillas are considerably larger than Australopithecines, whose stature is estimated to have been between 4 feet 6 inches and 5 feet 6 inches.

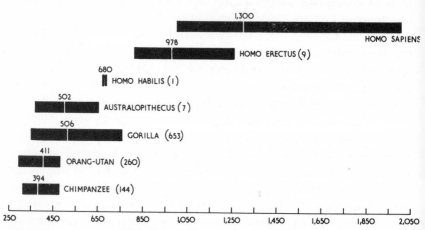

Figure 26. Cranial capacities of hominoids, the mean sample range and sample size are given for each group. The area for *Australopithecus* represents the estimated population range. Mean ± 3 SD. (Adapted from Figure 11, by P. V. Tobias, *Olduvai Gorge*, volume 2, Cambridge University Press, 1967.)

(2) Occipital condyles, the processes which articulate the skull with the first vertebra of the spinal column, lie farther forward on the skull base than they do in apes, but not quite so far forward as in modern man. The position of the condyles reflects the degree to which the skull is balanced on the vertebral column. Apes having heavy faces and jaws and backwardly placed occipital condyles require enormously strong muscles at the back of the neck to maintain the skull in equilibrium; modern man having relatively light jaws and a centrally placed fulcrum needs much

lighter muscles (Figure 20). Australopithecines are more man-like in this respect than ape-like (Figure 27).

(3) The projection of the face is moderate in extent showing neither

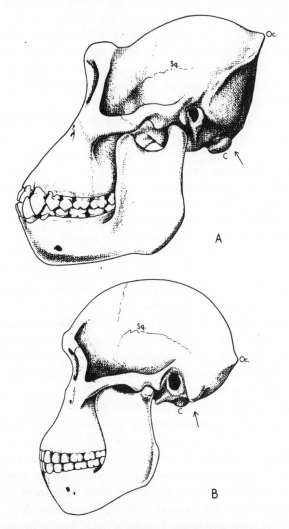

Figure 27. The skull of (A) female gorilla compared with (B) an australo-pithecine skull *(A. africanus)*. Note the relative positions of occipital condyle (C) and the occipital protuberance (Oc). Note also the diminished jaw protrusion in B and the absence of projecting canines. (From *Man-Apes or Ape-Men*, by W. E. Le Gros Clark. Courtesy of Holt, Rinehart & Winston, 1967.)

the heavy prognathous muzzle of the apes nor the straight (orthognathous) flat face of modern man.

(4) The tempero-mandibular joint, by which the lower jaw articulates with the skull, is human-like in its configuration and not a bit ape-like. The mechanics of the human joint differ from that of apes inasmuch as rotatory movement of the jaws, which provide a side-to-side or "gum-chewing" action, are possible. The large canines of chimpanzees and gorillas effectively block side-to-side movements of the jaws. Australopithecines having short canines were not so impeded and could have chewed gum with the rest of us.

(5) The teeth are arranged in a parabolic curve. The molar teeth are of moderate size and their cusps are rounded; the premolars are homomorphic (having the same shape). The canine tooth is small and incisor-like and does not project below the level of the toothrow, and the incisor teeth are broad and spatulate in the upper jaw. These characters are essentially similar to those of man. The differences, such as they are, can be dismissed as fairly trivial. The dentition of apes on the other hand is strikingly different. The palate is rectangular in its overall shape. The molars are extremely large and usually increase in size from the first to last molar. The molar cusps are rather sharp and conical before wear occurs and the surface enamel may be extremely wrinkled as in the orang-utan or moderately wrinkled as in the chimpanzee. The premolars are hetermorph (having different shapes); the first premolar in the lower jaw is especially modified to shear against the long projecting canine of the upper jaw and is described by anatomists as *sectorial*. The canines are very large teeth, conical in shape, projecting well below and above the occlusal surface of the toothrow. When the jaw is closed the canines of apes interlock, the lower canine fitting into a space in the upper jaw called the *diastema*, which is sited between the canine and lateral incisor.

(6) Among the parts of the skeleton that are known, the pelvic bones are perhaps the most revealing indicators of the true status of *Australopithecus*. Figure 18 shows the striking differences in shape of the pelvis in apes (in this instance a gorilla) and man. Note the great length of the portion *above* the hip joint (the ilium) in gorillas compared with man. Note, too, that the portion of bone *below* the hip joint (the ischium) is much shorter in man than in gorillas. Next, observe how the ilium of the gorilla faces backwards, while that of man faces sideways. These are a few of the obvious differences, but there are many

more less obvious ones. The pelvis of australopithecines is intermediate. This doesn't say much; it merely means "in between." Let us assume that A = *Australopithecus,* H = *Homo,* G = *Gorilla;* then "G . . . . . A . . H" provides a measure of the relative affinities of the australopithecine pelvis. In functional terms, the imperfections of the australopithecine pelvis (relative to man) indicate that the bipedal gait of *Australopithecus* was less well evolved than in modern man. For one thing, striding was almost certainly not present. Lacking a stride, *Australopithecus* would have walked quite effectively and efficiently, but without the drive and physiological economy that characterizes the striding gait of man. His walk would have been more of a jog-trot. As S. L. Washburn has said, "Man ran before he was able to walk." If we equate "walking" with "striding" then this is a fair summary of the australopithecine gait.

So far we have discussed only fossil material that relates to the genus *Australopithecus.* This genus, however, is not the only austra-lopithecine genus to be found in South Africa. To be considered is another hominid which has been given the scientific name of *Paran-thropus* Broom, 1954. At this point, we are entering into an extremely controversial field. Few authorities accept the interpretation that the two forms found at South African fossil sites are sufficiently different to justify generic separation. Dr. J. T. Robinson, late of the Transvaal Museum, Pretoria, now at the University of Wisconsin, whose familiarity with the australopithecine material is unrivaled, is the leading prota-gonist for the separationist school (1963). Other authorities such as Professor Bryan Patterson of Harvard and Dr. C. K. Brain of the Trans-vaal Museum are of a like mind. The present author has been *Paran-thropus*-minded since 1962.

Paranthropus is a genus of fossil hominids found in South Africa at Swartkrans and Kromdraai and in East Africa at Olduvai, and at nearby Peninj, Lake Natron. Regarded by most authorities as either a species of *Australopithecus,* or at best a subgenus, *Paranthropus* nevertheless differs significantly in the characteristics of its dentition and skull from *Australopithecus;* but it is in the anatomical characteristics of pelvis and femur that *Paranthropus* shows such striking distinctions (Napier, 1964; Day, 1969). While *Paranthropus* was undoubtedly bipedal in the broadest sense of the word, the gait was inefficient and clumsy; indeed this creature may well have resorted to a form of quadrupedalism (perhaps even knuckle-walking) under duress.

Robinson's analysis of skull and dental characters combined with

various observations on the postcranial material, point to a major difference in ecological background for *Paranthropus* and *Australopithecus*.
The following hypothesis would account for such differences in structure and behavior as have been inferred for *Paranthropus*. Arising from
a common arboreal hominoid stock during the early Miocene (Figure 17)
the ancestral hominids came down to the forest floor, while the African
pongids continued their largely arboreal existence until the late Miocene
or early Pliocene, when they too, for reasons possibly related to size, for-
sook the trees in favor of life on the ground. The hominid stock eventually moved out of the forests into woodland savanna. During the
Pliocene a dichotomy occurred amongst the hominids; part of the population (Australopithecus-to-be) migrated into the open savannas while
the other part (Paranthropus-to-be) stayed in a woodland habitat; and
there they remained—ecologically bolstered but evolutionarily stagnant—
until the late Pliocene or early Pleistocene when shrinking forests and
woodlands and perhaps competition from other primates, drove them
out into savannas and grasslands. There is little doubt that *Paranthropus*
and *Australopithecus* were living side by side throughout the early
Pleistocene in Africa but, being ecologically disparate, they were not in
serious competition with each other (Figure 28). Climatic evidence from
geological deposits in South Africa suggests that during the early Pleistocene, periods of dryness were interspersed with moister periods in which
rainfall maxima were approximately 40 inches per year and when a much
more densely vegetated habitat than that which exists today would have
prevailed. Robinson has suggested that *Paranthropus* was a vegetarian on
the basis of the proportions and wear of his teeth. As such he could have
subsisted on a diet of leaves, fruits, seeds, ground plants, roots, tubers,
and nuts of various kinds. *Australopithecus* in South Africa, like *Homo
habilis* at Olduvai Gorge in East Africa, was a primitive hunter and also
a scavenger, pre-empting carnivore kills, scaring off lions and leopards by
such crude but effective techniques as stone throwing and shouting. Thus,
without the threat of competition for food, the carnivore and the herbi-
vore could have lived side by side just as gorillas, chimpanzees, and man
do today in the forests of central Africa.

Paranthropus was a by-product of human evolution, a sort of drop-
out, like the gorilla, a species that preferred the placid, nonaggressive,
laissez-faire life of the forest to the rat race of the plains. *Paranthropus*
survived well up into the middle Pleistocene. For three quarters of a
million years this creature, originally called "Zinjanthropus" by its dis-

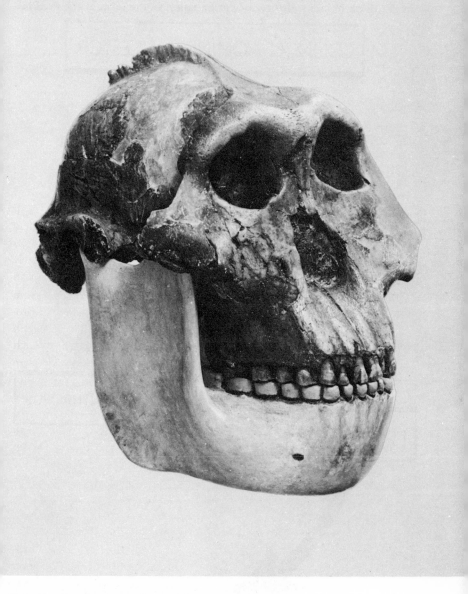

Plate 17.   Reconstruction of skull of "Zinjanthropus" from Bed I Olduvai
Gorge. (Photography by R. Campbell and A. R. Hewes, courtesy of
Cambridge University Press.)

coverer (Leakey, 1959) was a contemporary of *Homo habilis* but eventually he became extinct and left no descendants behind except, one might think in moments of unscientific fancy, in the form of the mythical

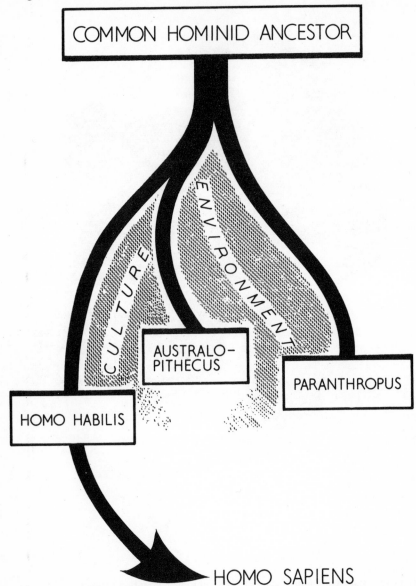

Figure 28. Author's concept of the relationship between (and the factors that separated) the early Pleistocene hominids.

hairy men of the woods, the Abominable Snowman, the Bigfoot, and the rest of that ilk. *Paranthropus* is memorable chiefly as counterpoint to the main stream of human evolution represented by *Australopithecus* and *Homo habilis*—tool-using, meat-eating, weapon-toting, regular guys on their way up.

Still, philosophically, within the area of major controversy but displaced, geographically, some 1000 miles to the northeast, we find ourselves in Olduvai Gorge, Tanzania—in Leakey Country. The early Pleistocene occupants of the lake shore sites that are now Olduvai Gorge were possibly the earliest known species of mankind, whose roots this book has been all about. In this sense we are nearing the end of the trail.

Olduvai Gorge, 25 miles long and 300 feet deep, cut from the

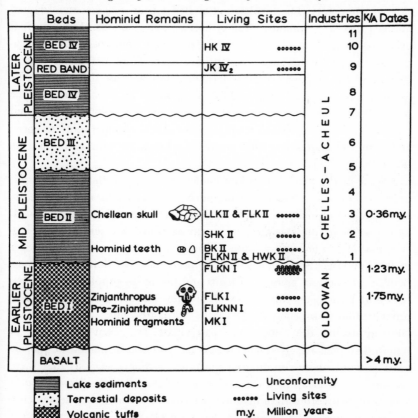

Figure 29. Stratigraphy of hominid remains and living sites, and potassium-argon dates in Beds I-IV at Olduvai Gorge. (From *The Prehistory of East Africa*, by Sonia Cole. Courtesy of The Macmillan Co., New York, 1963.)

Serengeti Plain in recent times by seasonal torrents, is a latter-day Grand Canyon. What it lacks in length, depth, and antiquity it gains in the richness of the contents of its rocks. The profile of the canyon is comprised of a series of geological beds exposed by weather erosion (Figure 29). The beds consist of layers of bentonitic clay derived from still-water lake deposits and interspersed with volcanic ash; the whole stratigraphic sequence being bedded on a thick flow of volcanic basalt. The beds are numbered I to V; the lowest—and therefore the oldest—is Bed I. It was here, some 24 feet below its junction with Bed II, that the first specimen of *Homo habilis* was discovered by the Leakeys in 1960-1961. The indications from this and other sites in Bed I that have yielded *Homo habilis* remains are that these early hominids were occupying temporary lakeside encampments. The fossil fauna of Bed I consist of savanna-adapted animals such as elephants, giraffes, hyenas, and baboons. The principal evidence for the way of life of early man comes from the study of occupation sites or "living floors" which have been found at five horizons in Bed I and two horizons in Bed II (Figure 29). At least three out of the five horizons in Bed I have yielded up remains of *Homo habilis*. The living floors are covered with fragments of fish, reptile, bird, and mammalian bones as well as manufactured pebble chopper-tools, scrapers, and primitive hand-axes, and unworked chunks of lava and quartzite (Plate 18).

Some of the mammalian long bones, many of which had been split or cracked open presumably for the purpose of extracting marrow, clearly belong to big animals, probably bovids. There were also bones of small animals—young pigs, rodents, and tortoises. This "kitchen midden" suggests that *Homo habilis* was both a hunter of small game and a scavenger of flesh of larger mammals killed by predators. No doubt he also depended on a certain amount of fruits and berries to supplement his diet. It would have been this aspect of his life, however, that might have brought him into competition with "Zinjanthropus," also found at Olduvai during the same time span (Plate 17). That two forms of hominid *could* have lived side-by-side for a period estimated at 750,000 years without one exterminating the other, strongly suggests that there was minimal overlap in their food requirements. Perhaps *Homo habilis* was more carnivorous than his successor *Homo erectus* whose dependence on vegetarian elements in the diet has been placed as high as 75 percent (Howell and Clark, 1963). This coexistence at least provides confirmation of the view of Robinson and others that *Paranthropus*, which includes "Zinjanthropus," was a vegetarian. It has recently been suggested by Clifford Jolly of New York University that *Paranthropus* was "graminivorous," subsisting on small, tough items, such as seeds, stems, and rhizomes. One cannot but think that *Paranthropus'* doom would have been sealed at a much earlier date if he had dared to be anything else than vegetarian. Some authors (e.g., Buettner-Janusch, 1969) have suggested that a species of *Paranthropus* and a species of *Homo* were in fact never contemporaneous at Olduvai Gorge. If they were not, it is difficult to explain the *mixed* nature of the hominid fossils at two of the "living sites" in Bed I that contain remains of both.

Man the Toolmaker At all levels of Bed I and Bed II there is unequivocal evidence of toolmaking. According to Mary Leakey (1966), pebble-choppers, the most characteristic artifact to be found, consist of a cobblestone which has been crudely flaked by three or four strokes on two opposing faces of one end or one side to produce a chopping edge; the unworked rounded butt fits naturally into the palm of the hand. In addition to lava pebble-choppers, spherical balls of quartzite, quartz scrapers, quartz flakes, and a crude type of hand-ax were also present on these living sites. Significantly, the hand-axes do not occur in the lower part of Bed I, but become increasingly common at the higher levels of Bed I and the lower levels of Bed II. It appears as if this new form of tool, characteristic of later, middle Pleistocene cultures, was in the "developmental" stage but not yet in "production" at Bed I Olduvai Gorge. The element of craftsmanship in which manual delicacy and precision play a large part was generally speaking absent from the Oldowan culture. Nevertheless neither the pebble tools, nor any of the other variants can be regarded as mere accidents; they show, in the constancy of their shape, that an industry had been developed in which tradition and learning were playing their part.

Ad hoc tool-using, in the sense of an improvised act employing a naturally occurring object such as a stick or a stone, has been observed in many non-human primates in the laboratory and in the wild. Mammals like the sea otter and birds like the Galapagos woodpecker-finch and even some invertebrates, use tools in the broadest sense of the word. A more purposeful form of tool-using is that shown by some wild chimpanzees which select stems and stalks of suitable size for the particular purpose of using them as "fishing rods," which are then poked into termite hills. When the sticks are withdrawn they are covered with termites which the chimpanzee proceeds to pick off with its lips. This type of purposeful tool-use was first seen by Jane van Lawick-Goodall during her study of wild chimpanzees in the Gombe Stream National Park in Tanzania.

Jane van Lawick-Goodall's chimpanzees even went one stage further, to a stage that I have called "tool-modifying." When the twigs they had plucked for their termite fishing proved to have awkward side branches, they would be stripped off before being put to use. Under experimental conditions in captivity chimpanzees will "modify" a tool by fitting one hollow bamboo cane into another in order to obtain sufficient length to reach a food reward. Tool-modifying involves a considerable element of conceptualization. The chimpanzee in other words is required to *think ahead.*

This ability of wild chimpanzees clearly suggests how toolmaking in man might have evolved from ad hoc tool-using, purposeful tool-using, and tool-modifying. It is but a short step from modifying a tool for an immediate purpose to modifying a tool for a future eventuality which, after all, is the essence of toolmaking. Although *Australopithecus* was not a toolmaker in a cultural sense, as was *Homo habilis,* he was undoubtedly an advanced tool-modifier. According to Professor Raymond Dart, the principal material of *Australopithecus* was not stone but bone. Dart has given the "industry" the jaw-cracking name of the osteodontokeratic culture. It is doubtful whether it can be called an industry in quite the sense that the term is defined below, nevertheless its reality is beyond doubt.

The culminating event in this sequence from tool-using to toolmaking was the advent of a toolmaking industry or the "culturization" of toolmaking. The archeological characteristic of a toolmaking culture is, as Louis Leakey has expressed it: "tools made to a set and regular pattern." The essential ingredient of cultural toolmaking is the element of tradition by which a skill is passed down from one generation to another, and it is possible that evolution of cultural toolmaking, the final stage of the toolmaking saga, followed close on the heels of the evolution of speech and language. Speech, however, is not the only means by which information can be passed from one individual to another; learning by example would be a perfectly feasible way in which one hominid could teach another an activity as simple as pebble-tool making. Indeed it has never seemed to me that lack of language and speech would be any bar to the development of simple cultural traditions, although it has for long been implicit that speech and culture are so reciprocally linked that one is not possible without the other. I would agree with S. L. Washburn (1960) when he said that "language may have appeared together with fine tools, fire and the complex hunting of large-brained men of the middle Pleistocene. . . ."

Toolmaking was technologically in its infancy in Bed I at Olduvai Gorge. The hand-ax culture, which succeeded it, occurred in two phases. The older phase, known as the Chellean, in which the product was a crude, roughly oval stone and the more recent—the Acheulean—a much more delicately flaked, pear-shaped stone, decorated by scalloping with sharp edges. The pebble-chopper was probably the cultural forerunner of the hand-ax of Europe and Africa and of the crude chopper of the Far East. It is interesting to speculate why and how the toolmaking culture spread from Africa. Clearly it must have spread while it was still at the pebble-chopper stage. The assumption is that man, fresh in from Africa, spread out from this general area of Europe towards the west where the hand-ax culture slowly evolved, and to the east where the chopper culture subsequently developed. Hand-axes have never been found further east than India and choppers of the type found at Choutkoutien, China, the home of Pekin Man (*Homo erectus*), have never turned up in the West. It seems that two cultures existed even then.

Although pebble-choppers were crude they enjoyed a phenomenal success—they lasted for nearly 1.5 million years before being superseded. This probably reflects the evolutionary inertia of the early stages of hominine evolution rather than the efficiency of pebble-choppers. During a period of 1.5 million years the cranial capacity of the brain increased by 32 percent from 680 cc (*H. habilis*) to 900 cc (*H. erectus*). During the next 250,000 years, or even less if the recent discovery of the back part of a skull from Vertesszöllös (Thoma, 1966) in Hungary is taken into account, it increased from 900 cc (*H. erectus*) to 1325 cc (*H. sapiens*, Swanscombe Man), or by 47 percent. In other words, it would seem likely that the major intellectual advances of the genus *Homo* occurred at the *H. erectus* stage, accompanying progress in toolmaking, advances in big-game hunting techniques, the discovery of fire, and the development of speech and language.

Homo habilis at Olduvai   The first discovery of an early Pleistocene hominid at Olduvai Gorge took place in 1959. "Zinjanthropus" or "Nutcracker Man," so-called because of his extremely powerful jaws, was discovered by Mary Leakey, eroded out of the slope of the Gorge with only his premolar teeth showing. On excavation, the site proved to contain bones of animals and pebble-chopper tools, and for a brief shining moment "Zinj" enjoyed the fame of being the earliest known specimen of man-the-toolmaker. Subsequently, early in 1961 at a slightly lower geological level and a few hundred yards away, the skull and jaw of an adolescent hominine, some foot bones, a collar bone and some hand bones came to light. Provisionally this newcomer was referred to as "pre-Zinj." It soon became clear that "pre-Zinj" was very different from "Zinj." Zinjanthropus had all the hallmarks of *Paranthropus,* already well known from South African fossil sites at Kromdraai and Swartkrans. "Pre-Zinj" with his smaller teeth and more delicate skull was clearly a close relation of *Australopithecus.* As "pre-Zinj" was undoubtedly the more man-like form, the epithet "toolmaker" was withdrawn from "Zinj" and awarded to "pre-Zinj." "Pre-Zinj" was subquently given the formal name *Homo habilis* Leakey, Tobias, and Napier, 1964. (See discussion on *Homo habilis* controversy, p. 139 and below.) Still further specimens of both *Paranthropus* (now regarded as either *Australopithecus boisei, Australopithecus [Paranthropus] boisei,* or *Paranthropus boisei,* depending on which side of the taxonomic fence you happen to be sitting) and *Homo habilis* (or, as his detractors will have it, *Australopithecus habilis*—and his flatterers, *Homo erectus habilis*) have been found in the lower part of Bed II. The date arrived at by potassium/argon tests for rocks of middle Bed I is approximately 1.75 million years and the estimated date of Bed II *H. habilis* sites, is 1 million years. Thus early man was no flash-in-the-pan, but a native East African for at least 0.75 million years.

The following is a brief summary of the characters of the skeleton of *Homo habilis.* Differences from *Australopithecus* and modern man are indicated where relevant.

1. A well-rounded brain case showing curvatures which are more *Homo*-like than any australopithecine and having a cranial capacity of 680 cc (Tobias, 1964).

2. A lower jaw which is less heavily muscled and has a more rounded dental arcade than *Australopithecus;* unlike modern man, however, it lacks a chin.

3. The teeth of *Homo habilis* are smaller than in *Australopithecus* and the premolars lack the characteristic side-to-side broadening.

4. The bones of the hand reveal that the thumb was opposable and that the last bone of the thumb was almost as broad as it is in modern man; it is, thus, quite unlike the thumb of a living chimpanzee or

a gorilla. This bone is not known for *Australopithecus*. The *H. habilis* hand was probably rather stubby and the fingers powerfully muscled (Napier, 1962).

5. The collar bone of *Homo habilis* is surprisingly big compared with other parts of the skeleton but it very closely resembles in shape the collar bone of modern man.

6. The foot specimen (Davis, Day, and Napier, 1964), which is complete except for the ends of the toes and the back of the heel, had an arch system comparable to that of modern man and, like man, the big toe was extremely stout and was aligned alongside the short toes (Plate 19.)

Although the bones on which this description is principally based was a "youth," there is every reason to believe that adult *H. habilis* was quite small, not more than 4 feet to 4 feet 6 inches in height (Plate 20) There is no doubt that he was bipedal and walked very much as modern man does, only lacking perhaps the full refinements of the striding gait.[16]

Homo habilis had a capable hand (*habilis* in fact means capable) but just what it was capable of is difficult to interpret. I am sure that *H. habilis* had a strong power grip, but I doubt if he had yet acquired a full-fledged precision grip. Possibly his thumb (as certain features of the existing bones suggest) was still rather on the short side. This obviously invites several questions. Was the hand of *H. habilis* advanced enough to have made the tools found associated with his remains? Is it possible to make tools of this sort without being able to oppose the thumb to the tip of the index finger? Experiments in toolmaking by the writer proved it to be perfectly easy to make pebble choppers without using the thumb at all! So, the answer is probably "yes." These, and some other simple experiments along the same lines, suggested to me that there might be some relationship between the nature of the tools and the nature of the hands that made them. It is possible, in other words, that the long period of technological inertia, lasting over 1.5 million years when the pebble chopper and its simple variants were the only sort of *stone* tool that man possessed, was simply the result of a period of stagnation in the evolution of the full potential of the human hand.

---

[16] In 1961 from quite a separate site at the top of Bed I, and many thousands of years later in time, a *hominid* big-toe bone came to light (Day and Napier, 1966). It shows characteristics which assuredly indicate that the owner walked erect with a striding gait which, as discussed on page 171, depends principally on the characteristics of the big toe.

Since 1967 a new name, a strange detergent-sounding name, has come into the picture—the Omo Valley. The Omo is a river in southwest Ethiopia that empties into Lake Rudolf. Hominid remains have been recovered from twelve sites in the lower Omo Valley which on the basis of potassium-argon dating techniques range from 2 to 4 million years old. Here as a result of expeditions led by F. Clark Howell of the University of Chicago, Richard Leakey, son of Louis, of the National Museum of Kenya, Nairobi, and Professor Arambourg of France, certain of the gaps in the history of man in his pre-human stage promise to be filled in when the material consisting of several lower jaws and many isolated teeth has been fully studied. In the meantime a preliminary report by Clark Howell (1969) suggests that both *Australopithecus* and *Paranthropus* were present in the lower Omo Valley four million years ago. Thus, in a matter of ten years our ideas of the time scale of hominid evolution have been radically changed. In 1959 most of us would have settled for 1 million years as a reasonable estimate for the age of the Australopithecines. By the early 1960s there was scientific evidence that this period extended back in time for nearly 2 million years and now, in 1969, it has been doubled once again.

Higher in Bed II, almost at the top in fact, at a stratigraphic level dated by the potassium-argon method to 490,000 years ago, a specimen of *Homo erectus* (originally called "Chellean Man") was discovered by the Leakeys in 1960. The specimen consists of a skull cap which is flattened and has a prominent nuchal crest where the neck muscles are attached at the back, and large, protruding brow ridges. In every way the "Chellean" skull, as it has been called, looks like a slightly progressive cousin of Java Man and Pekin Man. The cranial capacity of the skull has been estimated by Professor Phillip Tobias of Johannesburg at 1000 cc which puts it well within the range of *Homo erectus*. Its position in Bed II makes it a feasible descendant of *Homo habilis* of Bed I and II. The existence of the *Homo erectus* grade of evolution in Southeast Asia (Java), East Asia (Pekin), Europe (Vertesszöllös), North Africa (Algeria), South Africa (Swartkrans), and East Africa (Olduvai) makes it clear that man had become a polytypic species 0.5 million years ago. In addition to *Homo habilis, Paranthropus boisei,* and *Homo erectus* there is still another discovery that must be mentioned. In 1939 Hans Reck unearthed a skeleton of a relatively recent burial of *Homo sapiens* in Bed V (the top-most Bed, not illustrated in Figure 29). Reck's man is the icing on top of the Olduvai layer cake.

This remarkable site is unequaled anywhere in the world for its contribution to our knowledge of human history. Its human occupants represent the three known species of *Homo*—*H. habilis, H. erectus, H. sapiens*—and between them they span nearly two million years of time. The story of human occupation at Olduvai is even more complete than its tally of occupants suggests. From the bottom to the top, there is an almost unbroken sequence of stone tools representing almost every recognized African cultural phase from the Oldowan to the Capsian, from the days of simple pebble tools to an era of advanced stone technology. Furthermore, as L. S. B. Leakey—the James Cook of Olduvai Gorge—has observed, there are 150 miles of exposed fossil-bearing deposits yet to be excavated.

Relationships Between Pleistocene Hominids During the Pleistocene which probably lasted for at least 3.5 million years according to the latest estimates, there were three species of near-man in existence and three species of true-man.

Near-man: *Australopithecus africanus, Paranthropus robustus, Paranthropus boisei.*

True-man: *Homo habilis, Homo erectus, Homo sapiens.*

How are they related to each other? Figures 17 and 28 express the author's view of the phylogeny of the hominids. It is not a particularly controversial one and accords, broadly speaking, with the views of many authorities of the "early" school except in the matter of the phyletic position of *Paranthropus*. In Figure 17, the australopithecine line is envisaged as diverging from the main line of the hominid stem in the middle-to-upper Pliocene. The members of the genus *Australopithecus* that we know from the fossil record in South Africa are not ancestral to later hominids—they are out of the running, a nonprogressive sideline. Nevertheless, they provide a reasonably close model for the ancestors of *Homo habilis*. As the line leading to *Homo* is the progressive one, the hypothetical common ancestor of both *Australopithecus* and *Homo habilis* probably looked more like the former than the latter. Behaviorally, such a creature would have been a biped (though not a very efficient one), a tool user (but not a toolmaker), a meat eater (but principally as a result of scavenging), a socially oriented being (but organized on the troop principle of the common baboon or, perhaps, on the harem principle of the hamadryas baboon and the gelada) and communicating (but not articulate).

The three lines that led to *Paranthropus, Australopithecus,* and *Homo* became separate, firstly, as a result of ecological factors and, secondly, because of the introduction of a behavioral or "cultural" factor—toolmaking (Figure 28).

It is a reasonable assumption that the ancestors of *Homo habilis* were Australopithecines, and that the Australopithecines, known from the South African fossil record in the lower Pleistocene, were also descendants of this original stock. In this sense, *Homo habilis* is an advanced australopithecine. From the evidence already adduced in this chapter, however, it seemed to Leakey et al. (1964) that *H. habilis* had progressed beyond the australopithecine grade and crossed the threshold of humanity. This threshold, by the nature of things, is very difficult to recognize within the limitations of current taxonomic procedure, but there is no doubt that a point of transition must have existed. It is entirely a matter of personal opinion as to exactly where the dividing line should be placed. As has been discussed elsewhere (p. 141), such *formes de passage* offer tremendous problems to taxonomists.

As a colleague of mine, Dr. Michael Day, has written (1965) "The situation [the *Homo habilis* controversy] has exposed the weaknesses of current systematics and points to the need of a classification which can accommodate fine gradations between forms, while allowing each a normal range of anatomical variation."

One might ask why anthropologists try to work within such an unwieldy framework; why they attempt slavishly to adhere to a Linnaean classification which is inadequate to cope either with *formes de passage* or with the ecological, behavioral, and cultural information that constitutes the principal evidence for progressive evolution of man in the Pleistocene. Why indeed?

It is ironical that some of the most interesting material from the Bed I deposits of *Homo habilis* is inadmissible as evidence for the taxonomic status of the species, e.g., the foot, the hand, and the collar bone. As there is no comparable material from australopithecine sites to compare these structures with, they cannot be used as evidence to show that *H. habilis* is indeed *Homo*. But, like the evidence of toolmaking, these bones are nevertheless pointers to the *real* biological nature of the creature that possessed them irrespective of taxonomic considerations which, in this instance, serve only to shackle the truth. Taking all factors into account there can be little doubt that *Homo habilis* is more hominized than

Plate 18. A section of a "living floor" at Olduvai Gorge collected by Professor P. R. Davis. Stone tools and large bovid bones can be seen.

Australopithecus and less hominized than *H. erectus* who, in turn, is less hominized then *H. sapiens*. Therefore, notwithstanding the systematic statutes and nomenclatorial niceties, we can recognize four *grades* of hominization within the Pleistocene: *Australopithecus, Homo habilis, Homo erectus*, and *Homo sapiens*. Each successive grade provides a substantial advance in structure and behavior over the preceding one. Even if these species were called John, Paul, George, and Ringo they would *still* form a series of morphological grades (Table 13).

Table 13 **African and Asian Representatives of Hominid Grades During the Pleistocene**

Grade	African	Asian
"John"	*Australopithecus africanus*	*Meganthropus paleojavanicus* (Java,
"Paul"	*Homo habilis* (Bed I, Olduvai)	Djetis Beds)
"George"	*Homo habilis* (Bed II, Olduvai)	*Homo erectus erectus* (Java, Trinil Beds)
"Ringo"	*Homo erectus* ("Chellean Man")	*Homo erectus pekinensis* (Choukoutien)

The assumption is generally made that Africa is the sole and original site of human origins. Darwin suggested it was so, and the discoveries in South Africa and Olduvai Gorge seem to have removed any lingering doubts there may have been on the matter. Perhaps, too, the unconscious wish to derive all mankind from a single geographical stock in the cause of racial togetherness, has been a contributory factor. Undoubtedly "races" existed at that time because races have always existed (the term is equivalent to subspecies which is a biological concept that defines potentially interbreeding populations of a single species separated by geographical barriers). Modern races of man, the delineation of which is still quite unclear, are a relatively recent phenomenon. Most authorities today look upon modern races as having arisen from a common ancestral *Homo sapiens* stock—from Cro-Magnon man perhaps—some 30,000 years ago. At this time, warmer postglacial conditions were opening up migratory routes into northern and eastern latitudes; this would have facilitated the dispersion of *Homo sapiens* and the beginning of the geographical isolation of populations necessary for the evolution of true races. No doubt, too, the development of agriculture and the establishment of settled communities hurried this process along; thus it is possible that the original races of mankind (whatever they were) are of an even more

Plate 19. Articulated foot bones of *Homo habili*

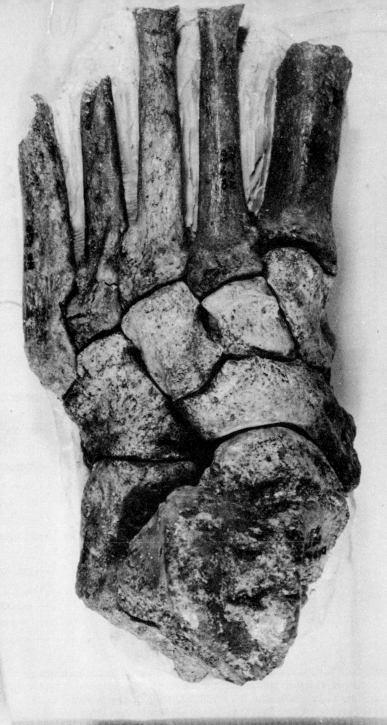

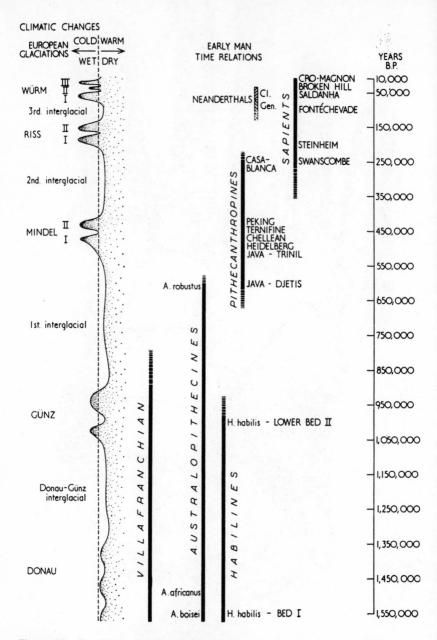

Figure 30. The Pleistocene epoch showing the principal fossils, their dates of occurrence, and their relationship with the European glaciations.
(Reproduced by permission of The World Publishing Company, from *A Guide to Fossil Man* by Michael Day. Copyright © 1965 by Michael Day.)

recent origin than we suppose—perhaps not more than 10,000 years old.

On the other hand Carleton Coon in his book *Origins of Races* (1961) extends the roots of the living races of man somewhat further back in time; he believes that races evolved when man was at the *Homo erectus* grade of evolution, some 0.5 million years ago. Coon believes that *Homo erectus* was transformed, asynchronously and in various parts of the world, into *Homo sapiens* not less than five times. Few anthropologists or geneticists today are willing to accept this hypothesis.

But to return to the problem of the birthplace of the genus *Homo*. Since the late Miocene there have been both hominids and pongids in Europe and Asia, as far west as Spain and as far east as China. Before that, arboreal gibbon-like hylobatids of the genus *Pliopithecus* were widespread in western Europe. It is the hominids, however, that interest us most. *Ramapithecus* (including *Kenyapithecus*) from the Siwalik Hills of India, Africa, and possibly from Europe too, is widely accepted as the late Miocene ancestor of man. The teeth and jaws, which are the only parts known, show quite striking resemblances to *Australopithecus* and to *Homo* and it seems quite possible that *Ramapithecus* is a linear ancestor of both these genera which subsequently replaced each other. A certain amount of zoogeographical juggling would seem to be required to get *Ramapithecus'* descendants back to Africa to be ancestors to *Australopithecus;* but in fact this is quite unnecessary. *Kenyapithecus* from Fort Ternan in East Africa, as we have already discussed, probably belongs in the same genus as *Ramapithecus*. Thus, this hominid could well have evolved into the African Australopithecines, while *Ramapithecus* was founding an Asian branch of the australopithecine dynasty on its own account. There is quite good evidence that hominid evolution was occurring, in parallel, in Africa and Asia in the early Pleistocene (Table 13). The existence of the australopithecine grade in Asia is uncertain but *Meganthropus paleojavanicus* from Java seems a likely candidate. To some authorities, *Meganthropus* seems to be the Asian equivalent of *Homo habilis* although somewhat more primitive in tooth structure. Considering the differing ecological situations extant in Asia and Africa during the early Pleistocene, it would be surprising to find that the grades were precisely similar. The specimen of *Homo habilis* (nicknamed "Cinderella") from lower Bed II, Olduvai Gorge has some striking similarities with an early Javan representative (formerly *Pithecanthropus* IV) of *Homo erectus* grade as Tobias and von Koenigswald have posited (1964). Finally the Bed II skull cap from Olduvai, provisionally called "Chellean

Man," is clearly a member of the *Homo erectus* grade and in terms of cranial capacity is on a par with the later Pekin pithecanthropine, *H. erectus pekinensis.*

The correlations between the eastern and western hominines are summarized in Table 13. It is clear that I am putting forward a case for the diphyletic origin of man up to and including the *Homo erectus* grade of evolution which arose independently out of an Asian *Ramapithecus* stock and an African *Kenyapithecus* stock. In other words I am suggesting that for a considerable period of time (ten million years or so) at least two groups of early hominids were evolving in parallel and in geographical isolation during the Pliocene and early Pleistocene.

However, I do not believe, as does Coon, that the *Homo sapiens* grade was achieved "not once but five times, as each subspecies, living in its own territory, passed a critical threshold from a more brutal to a more sapient state." *Homo sapiens,* modern man, whenever and wherever he arose, was the product of a gene pool which contained the genetical elements of both African and Asian hominid stocks. Vertesszöllös Man from Hungary with his large brain, advanced technology and culture, and "intermediate" geographical location seems to me as likely a candidate as we have at the present time for the role of either the earliest member of *Homo sapiens*—a monophyletic species world-wide in distribution.

The hominization process was going in Africa and in Asia and in Europe. We cannot be certain which continent started it all. If we go far enough back, it was probably Africa (Darwin right! Once again!) But which continent contributed most to the sum total of the gene pool of *Homo sapiens* is something that is beyond our ken and likely, for the moment, to remain so. Africa may have been the birthplace of the hominids and possibly of *Homo,* but there is no certainty that it was the sole nursery of modern man.

Plate 20. Author's reconstruction of *Homo habilis.* (Drawn by Barry Driscoll, courtesy of The Sunday Times, London.)

Finale

The central issue for anthropology today hinges on a single question. Is man's behavior simply a matter of history as Ortega y Gasset has averred, or is it, as Charles Darwin believed, the inevitable result of his zoological, non-human past? The straight answer lies not in the theorizing of amateur and professional social anthropologists but in the discoveries and interpretations of physical anthropologists and human paleontologists.

Man's ancestors are hominids, creatures who (as some, but not all of us, believe) have followed a separate evolutionary course for close to twenty million years. It is the behavior of man's direct ancestors that we must look to for illumination and not to the absorbing, but questionably relevant, antics of his primate cousins. A glance at Figure 4 reminds us that neither the ape stem nor the Old World monkey stem can offer any *direct* evidence of the prehuman behavior of hominids. As hominids are now represented by only a single species, *Homo sapiens,* we have no living hominid models from which inferences of proto-human behavior can be drawn and the only direct evidence we can hope for is that derived from fossils. The fossil record is sparse and our interpretative potential, particularly as regards the behavioral correlates of structure, is severely limited by the nature of the material. The living sites at Bed I, Olduvai Gorge, however, have shown that certain deductions can be made from the debris, the kitchen midden, associated with fossil bones. A new discipline of paleoarcheology may well be born to develop this aspect of paleoanthropology. Logically, therefore, paleoanthropology should receive the lion's share of financial support in the next decade for if we want to know more about the roots of mankind we must do a lot more digging (literally) into soils of the past.

The foregoing argument does not invalidate the zoological perspective of social anthropological thinking ably argued by Tiger and Fox (1966) and Fox (1967). The viewpoint of these authors, indeed, has been a principal stimulus for writing this book. It is simply an attempt to clarify what is not always clearly stated by popular writers on primate ethology that, for reasons apparent in the branching structure of the family tree, studies of primate behavior are teaching us more about primate behavior and evolution than about human behavior and human evolution. Fox (1967) is well aware of the inherent dangers of using non-human primate models. Primate behavior field studies have vastly increased our knowledge of how the lives of primates are organized, how the environment has influenced the direction of behavioral adaptation, and has put an edge on our awareness of the part played by culture in human adaptation. Failing the direct evidence of fossil hominids, man must gain what knowledge he may by analogy—by the indirect evidence provided by the study of non-human primates. Although the hominid stem has been separated from the ape stem for many millions of years, and from the Old World monkey stem for an even longer period of time, man shares a common inheritance with both apes and monkeys—a common genetic potential. Primates, particularly anthropoid primates, are, as I have indicated above, extremely conservative animals which have departed little from the ancestral pattern of primate structure and, thus, the subsequent evolution of non-human primates has run in close parallel with that of human primates. It is a perfectly reasonable proposition to study intensively the behavior of apes and monkeys for they provide the evidence, albeit indirect, that we lack for man. This is particularly relevant if the environments of the animals studied happen to correspond with the environment envisaged for early man. Thus ground-living primates such as baboons and macaques are probably more relevant to the human situation than the highly specialized forest-living chimpanzees and gorillas. Perhaps even more relevant are field studies of primates in transition from an arboreal mode to a terrestrial one. The savanna monkeys of Africa (*Cercopithecus aethiops*) provide the best example of this category. Savanna monkeys provide an evolutionary model for the metamorphosis of a species transferring from a forest to a savanna habitat, an ecological progression that, as we see it at present, recreates a critical phase in the story of our past.

In this short account of the roots of mankind I have tried to bring to your attention the genesis of man's physical characters and of some of his behavioral characters. I have attempted to show, principally in terms of structure, that the taproots of man are deeply entrenched in the soil of the past. Is man, man or beast? Man is both. He is both a good animal and a good man; and his future depends on his recognition of this duality in his nature. Man is a product of a primary arboreal background and a secondary ground-living heritage. He possesses the "flight" responses of forest-living monkeys and the "fight" reponses of the ground-living baboons and macaques. His genetic ambivalence confronts him every moment of the day. Do I get in there and fight, or do I settle for what I have? Do I "twist" or "buy"? Our uncertainty, the racking moments of self-doubt teetering on the precipice of indecision is a reflection, not of our ambivalent present, but of our ancient mixed-up past.

Self-immolation, however, can go too far. Konrad Lorenz, in a flash of pique, stated that "the missing link between the apes and *Homo sapiens* is Man." In moments of catharsis we tend to humble ourselves by exhibiting every last pucker of the scars that constitute our marks of Cain. This attitude we can understand but at the same time, intellectually, we reject it. Writers like Robert Ardrey, scientists like Professor Raymond Dart, have conditioned us to this form of self-abasement. But it could be just so much nonsense! E. O. Wilson, Professor of Zoology at Harvard, has expressed a note of optimism at the Man and Beast Symposium at the Smithsonian Institution in 1969, which may prevent the human herd of lemmings from hurling themselves into a watery oblivion of self-denigration. Why, he says, should we assume that the cloak of the early Pleistocene Elijahs, the aggressive killer-apes of Dart and Ardrey, has necessarily fallen upon us, the latter-day Elishas? Aggressiveness is largely a genetically determined phenomenon. On empirical grounds it is perfectly reasonable to assume that the genetic trait of agressiveness could have evolved in as few as ten generations, a matter of 300 or so years. Aggressiveness and the territorial imperatives of our society could simply be genetic adaptations to the demands of modern environments. "Original sin" in fact may be simply "adaptive sin," an evolutionary response to the environmental pressures of today. Our marks of Cain may simply be the scars of battle.

The Russells have said much the same thing in their book *"Violence in Monkeys and Man"* (1968). Violence, as they clearly demonstrate, is not some ultimate and irreducible feature of the universe, but a reaction of all mammals exposed to stress, the principal etiology of which is overcrowding. Violence and aggression are adaptive phenomena that serve to counter the extreme danger to the species of overpopulation. As the Russells say, if we can reduce overcrowding we can diminish violence. The remedy is there, but is mankind capable of taking the hint and accepting all that it entails in the way of compulsory birth control? Overpopulation may not lead to world starvation as Malthus prophesied; mankind is more likely to jostle itself to death. This is one of the lessons that has been learned from ethological studies of living primates.

The problem for behavioral scientists is no longer to convince other scientists that human behavior possesses biological components that obey the laws of Darwinian selection (few anthropologists or even sociologists, in their heart of hearts, would deny it any longer), but to demonstrate *just what these components are.* Above all, the function of behavioral scientists, indeed of anthropologists, anatomists, geneticists and the rest of us, is to convince those who walk the corridors of power of the truth of Earnest Hooton's plagiarism of the poet Terence: *"Primatus sum nihil primatus mihi alienum puto."* (I am a primate; therefore, nothing about primates is beyond my concern.) The power of corridors is that they provide easy access to the office of State and this is where the buck must reach before it stops.

The real significance of the study of monkeys and apes is that nonhuman primates provide a mirror into which man may look to discover what being a man is all about. But let us look at our reflected image critically and not emotionally; empirically and not intuitively; with reason and not with prejudice.

Literature cited

ANKEL, F.
  1965. Der Canalis sacralis als Indikator für die Länge der Caudalregion der Primaten. *Folia Primatologica,* 3:263–276.

ASHTON, E. H., AND C. E. OXNARD
  1964. Functional Adaptations of the Primate Shoulder. *Proceedings of the Zoological Society of London,* 142:49–66.

AVIS, VIRGINIA
  1962. Brachiation: The Crucial Issue for Man's Ancestry. *Southwestern Journal of Anthropology,* 18:119–148.

BARTH, FREDERIK
  1950. On the Relationships of Early Primates. *American Journal of Physical Anthropology,* 8 (2): 139–150.

BEARD, J. S.
  1953. The Savanna Vegetation of North Tropical America. *Ecological Monographs,* 23:149–215.

BERNSTEIN, I. S.
  1968. The Lutong of Kuala Selangor. *Behaviour,* 32:1–16.

BLURTON-JONES, N.G.
  1967. An Ethological Study of Some Aspects of Social Behaviour of Children in Nursery School. In D. Morris, editor, *Primate Ethology.* London: Weidenfeld and Nicolson.

BOURLIÈRE, F.; M. BERTRAND; AND C. HUNKELER
  1969. L'Ecologie de la Mone de Lowe (*Cercopithecus campbelli lowei*) en côte d'Ivoire. *La Terre et La Vie,* Number 2:135–163.

222

BRACE, C. L., AND M. F. ASHLEY MONTAGU
 1965. *Man's Evolution.* London: Collier-Macmillan.

BUETTNER-JANUSCH, J.
 1966. *Origins of Man.* Chichester: John Wiley.

BUTLER, H.
 1967. Seasonal Breeding of the Senegal Galago in the Nuba Mountains, Republic of the Sudan. *Folia primatologica,* 5:165–175.

CAIN, A. J.
 1960. *Animal Species and Their Evolution.* London: Hutchinson (1966).

CARPENTER, C. R.
 1940. A Field Study in Siam of the Behavior and Social Relations of the Gibbon *(Hylobates lar). Comparative Psychological Monographs,* 16:1–212.

CHALMERS, N. R.
 1968. Group Composition, Ecology and Daily Activities of Free-living Mangabeys in Uganda. *Folia primatologica,* 8:247–262.

CLARK, W. E. LE GROS
 1959. *The Antecedents of Man.* Edinburgh University Press.

CLARK HOWELL, F.
 1969. Remains of Hominidae from Pliocene/Pleistocene Formations in the Lower Omo Basin, Ethiopia. *Nature,* 223:1234–1239.

CLARK HOWELL, F., AND J. D. CLARK
 1963. Acheulian Hunter-gatherers of Sub-Saharan Africa. In F. C. Howell and F. Bourlière, editors, *African Ecology and Human Evolution.* Chicago: Aldine Publishing Company.

COLE, J.
 1963. *Macaca nemestrina* Studied in Captivity. In J. R. Napier and N. A. Barnicot, editors, *The Primates. Symposia of the Zoological Society of London,* Number 10:105–114.

COON, CARELTON S.
 1961. *The Origin of Races.* London: Jonathan Cape. (1963).

COWGILL, URSULA M.
 1970. The People of York: 1528–1812. *Scientific American,* 222:104–112.

CROOK, J. H.
 1966. Gelada Baboon Herd Structure and Movement. In *Symposia of the Zoological Society of London,* Number 18:237–258.

DAVIS, P. R.; M. H. DAY; AND J. R. NAPIER
 1964. Hominid Fossils from Bed I, Olduvai Gorge, Tanganyika. *Nature,* 201:967–970.

DAY, M. H.
 1965. Comment on "New Discoveries in Tanganyika: Their Bearing on Hominid Evolution" by P. V. Tobias. *Current Anthropology,* 6:391–399.
 1967. Olduvai Hominid 10: A Multivariate Analysis. *Nature,* 215: 323–324.
 1969. Femoral Fragment of a Robust Australopithecine from Olduvai Gorge, Tanzania. *Nature,* 221:230–233.

DAY, M. H. AND J. R. NAPIER
    1966. A Hominid Toe-Bone from Bed I, Olduvai Gorge, Tanzania. *Nature*, 211:929—930.

DE BEER, GAVIN
    1963. *Charles Darwin: Evolution by Natural Selection.* London: Nelson.

DOBZHANSKY, T.
    1963. Genetic Entities and Hominid Evolution. In S. L. Washburn, editor, *Classification and Human Evolution.* London: Methuen (1964).

EISELEY, LOREN C.
    1956. Charles Darwin. *Scientific American*, 194:62—72.
    1959. Charles Lyell. *Scientific American*, 201:98—106.

ELLEFSON, JOHN O.
    1968. Territorial Behavior in the Common White-Handed Gibbon. In Phyllis C. Jay, editor, *Primates.* New York: Holt, Rinehart & Winston.

FOX, ROBIN
    1967. In the Beginning: Aspects of Hominid Behavioural Evolution. *Man*, 2:415–433.

FREEDMAN, D. G.
    1967. A Biological View of Man's Social Behavior. In *Social Behavior from Fish to Man* by William Etkin. London: University of Chicago Press (1968).

GARDNER, R. ALLEN, AND BEATRICE T. GARDNER
    1969. Teaching Sign Language to a Chimpanzee. *Science*, 165:664–672.

GARTLAN, J. S.
    n.d. In J. R. and P. H. Napier, editors, *The Old World Monkeys.* New York: Academic Press [in press].

GAUTIER-HION, ANNE
    1966. L'Écologie et L'éthologie du Talapoin *(Miopithecus talapoin talapoin)*. *Biologica Gabonica*, 2(4): 311–329.

GOODMAN, M.
    1963. Man's Place in the Phylogeny of the Primates as Reflected in Serum Proteins. In S. L. Washburn, editor, *Classification and Human Evolution.* London: Methuen (1964).

GROVES, COLIN P.
    1967. Ecology and Taxonomy of the Gorilla. *Nature*, 213:890–893.

HARRISON, R. J., AND WILLIAM MONTAGNA
    1969. *Man.* New York: Appleton-Century-Crofts, Inc.

HEWES, G. W.
    1961. Food Transport and the Origins of Human Bipedalism. *American Anthropologist*, 63:687—710.

HILL, W. C. OSMAN
    1952. The External Genitalia of the Female Chimpanzee; with Observations on the Mammary Apparatus. *Proceedings of the Zoological Society of London*, 121:133–145.

JAY, PHYLLIS
    1965. The Common Langur of North India. In I. DeVore, editor, *Primate Behaviour.* New York: Holt, Rinehart & Winston.

JOLLY (NÉE BISHOP), ALISON
  1964. Use of the Hand in Lower Primates. In J. Buettner-Janusch, editor, *Evolutionary and Genetic Biology of Primates.* New York and London: Academic Press.
  1967. *Lemur Behavior.* University of Chicago Press.

JONES, MARVIN L.
  1962. Mammals in Captivity—Primate Longevity. *Laboratory Primate Newsletter,* 1 (3) :3–13.

JONES, F. WOOD
  1918. *Arboreal Man.* London: Edward Arnold.

LEAKEY, L. S. B.
  1959. A New Fossil Skull from Olduvai. *Nature,* 184:491–493.

LEAKEY, L. S. B.; P. V. TOBIAS; AND J. R. NAPIER
  1964. A New Species of the Genus *Homo* from Olduvai Gorge. *Nature,* 202:3–9.

LEAKEY, MARY D.
  1966. A Review of the Oldowan Culture from Olduvai Gorge, Tanzania. *Nature,* 210:462–466.

LEWIS, O. J.
  1969. The Hominoid Wrist Joint. *American Journal of Physical Anthropology,* 30:251–267.

LYELL, CHARLES
  1830. *Principles of Geology.* London: John Murray.

MACALPINE, I., AND R. HUNTER
  1969. Porphyria and King George III. *Scientific American,* 221 (1) :38–46.

MARTIN, R. D.
  1966. Treeshrews: Unique Reproductive Mechanism of Systematic Importance. *Science,* 152:1402–1404.
  1967. Towards A New Definition of Primates. *Man,* 2:377–401.

MAYR, ERNST
  1940. Speciation Phenomena in Birds. *American Naturalist,* 74:249–278.

MORRIS, DESMOND
  1967. *The Naked Ape.* London: Jonathan Cape.

NAPIER, J. R.
  1956. The Prehensile Movements of the Human Hand. *Journal of Bone and Joint Surgery,* 38B:902–913.
  1960. Studies of the Hands of Living Primates. *Proceedings of the Zoological Society of London,* 134:647–657.
  1962. Fossil Hand Bones from Olduvai Gorge. *Nature,* 196:409–411.
  1963. The Locomotor Function of Hominids. In S. L. Washburn, editor, *Classification and Human Evolution.* London: Methuen (1964).
  1964. Evolution of Bipedal Walking in the Hominids. *Archives de Biologie (Liege),* 75 (Supplement) :673–708.
  1967. The Antiquity of Human Walking. *Scientific American,* 216 (4) :56–66.

NAPIER, J. R., AND P. R. DAVIS
  1959. The Forelimb Skeleton and Associated Remains of *Proconsul africanus.* *Fossil Mammals of Africa,* Number 16. London: British Museum (Natural History) .

NAPIER, J. R., AND P. H. NAPIER
   1967. *A Handbook of Living Primates.* London and New York: Academic Press.

NAPIER, J. R., AND A. C. WALKER
   1967. Vertical Clinging and Leaping: A Newly Recognized Category of Locomotor Behaviour of Primates. *Folia primatologica,* 6:204–219.

PETTER-ROUSSEAUX, A.
   1964. Reproductive Physiology and Behavior of the Lemuroidea. In J. Buettner-Janusch, editor, *Evolutionary and Genetic Biology of Primates.* New York and London: Academic Press.

PILBEAM, D. R.
   1968. The Earliest Hominids. *Nature,* 219:1335–1338.

READ, DWIGHT W., AND PETE E. LESTREL
   1970. Hominid phylogeny and immunology. *Science,* 168:578–580.

RENSCH, B.
   1959. *Evolution Above the Species Level.* London: Methuen.

REYNOLDS, VERNON
   1967. *The Apes.* London: Cassell (1968).

REYNOLDS, VERNON, AND FRANCES REYNOLDS
   1965. Chimpanzees of the Budongo Forest. In I. DeVore, editor, *Primate Behavior.* New York: Holt, Rinehart & Winston.

RIOPELLE, A. J.
   1963. Growth and Behavioural Changes in Chimpanzees. *Zeitschrift für Morphologie und Anthropologie,* 53:53–61.

RIPLEY, SUZANNE
   1967. The Leaping of Langurs. *American Journal of Physical Anthropology,* 26:149–170.

ROBINSON, J. T.
   1963. Adaptive Radiation of the Australopithecines and Origin of Man. In F. C. Howell and F. Bourlière, editors, *African Ecology and Evolution.* Chicago: Aldine.
   1965. *Homo "habilis"* and the Australopithecines. *Nature,* 205:121–124.

RUMBAUGH, D. M.
   1965. Maternal Care in Relation to Infant Behavior in the Squirrel Monkey. *Psychological Reports,* 16:171–176.

RUMFORD, COUNT
   1802. Report Relative to the Present State of the Institution . . . 26th April 1802. *Journals of the Royal Institution,* 1:113–119.

RUSSELL, CLAIRE, AND W. M. S. RUSSELL
   1968. *Violence in Monkeys and Man.* London: Macmillan.

SARICH, V. M.
   1968. The Origins of Hominids. In S. L. Washburn and P. S. Jay, editors, *Perspectives on Human Evolution I.* New York: Holt, Rinehart & Winston.

SCHALLER, G. B.
   1963. *The Mountain Gorilla: Ecology and Behavior.* London: University of Chicago Press.

SCHENKEL, R., AND L. SCHENKEL-HULLIGER
   1967. On the Sociology of Free-ranging Colobus (*Colobus guereza caudatus* Thomas 1885). In D. Starck, R. Schneider and H. J. Kuhn, editors, *Progress in Primatology.* Stuttgart: Gustav Fischer.

SCHULTZ, A. H.
  1948. The Number of Young at Birth and the Number of Nipples in Primates. *American Journal of Physical Anthropology*, 6:1–24.
  1969. *The Life of Primates*. London: Wiedenfeld and Nicolson.

SIMONS, E. L.
  1967. Fossil Primates and the Evolution of Some Primate Locomotor Systems. *American Journal of Physical Anthropology*, 26:241–253.

SIMONS, E. L., AND P. E. ETTEL
  1970. *Gigantopithecus*. *Scientific American*, 222:76–85.

SIMPSON, G. G.
  1961. *Principles of Animal Taxonomy*. New York: Columbia University Press.
  1963. The Meaning of Taxonomic Statements. In S. L. Washburn, editor, *Classification and Human Evolution*. London: Methuen (1964).
  1969. *Biology and Man*. New York: Harcourt, Brace and World.

SUZUKI, A.
  1965. An Ecological Study of Wild Japanese Monkeys in Snowy Areas—Focused on Their Food Habits. *Primates*, 6:31–72.

SZALAY, FREDERICK S.
  1968. The Beginnings of Primates. *Evolution*, 22:19.

TANNER, J. M.
  1962. *Growth at Adolescence*. 2nd edition. Oxford, England: Blackwell.

THOMA, A.
  1966. L'Occipital de l'homme Mindelien de Verteszöllös. *L'Anthropologie*, 70: 495–534.

THORINGTON, RICHARD W. JR.
  1967. Feeding and Activity of *Cebus* and *Saimiri* in a Colombian Forest. In Starck, Schneider, and H. J. Kuhn, editors, *Progress in Primatology*. Stuttgart: Gustav Fischer.

TIGER, LIONEL
  1969. *Men in Groups*. London: Thomas Nelson.

TIGER, LIONEL, AND ROBIN FOX
  1966. The Zoological Perspective in Social Science. *Man*, 1:75–81.

TINBERGEN, N.
  1951. *The Study of Instinct*. Oxford University Press.

TOBIAS, P. V.
  1964. The Olduvai Bed I Hominine with Special Reference to Its Cranial Capacity. *Nature*, 197:743–746.
  1965. Early Man in East Africa. *Science*, 149:22–33.

TOBIAS, P. V., AND G. H. R. VON KOENINGSWALD
  1964. A Comparison Between the Olduvai Hominines and Those of Java and Some Implications for Hominid Phylogeny. *Nature*, 204:515–518.

TUTTLE, RUSSELL H.
  1967. Knuckle-walking and the Evolution of Hominoid Hands. *American Journal of Physical Anthropology*, 26:171–206.
  1969. Knuckle-Walking and the Problem of Human Origins. *Science*, 166:953–961.

VAN LAWICK-GOODALL, JANE
    1969. The Behaviour of Free-living Chimpanzees on the Gombe Stream Reserve. *Animal Behaviour Monographs*, 1(3):161–311.

WALKER, ALAN
    1967. Patterns of Extinction Among Madagascan Subfossil Lemuroids. In P. S. Martin and H. E. Wright, Jr., editors, *Pleistocene Extinctions*. Yale University Press.

WALKER, ALAN, AND M. D. ROSE
    1968. Fossil Hominoid Vertebra from the Miocene of Uganda. *Nature*, 217:980–981.

WASHBURN, S. L.
    1950. The Analysis of Primate Evolution with Particular Reference to the Origin of Man. *Cold Spring Harbor Symposia of Quarterly Biology*, 15:67.
    1960. Tools and Human Evolution. *Scientific American*, 203:63–75.
    1967. Behavior and the Origins of Man. Huxley Memorial Lecture. *Proceedings of the Royal Anthropological Institute*, 3:21–27.

WASHBURN, S. L.; PHYLLIS C. JAY; AND JANE B. LANCASTER
    1965. Field Studies of Old World Monkeys and Apes. *Science*, 150:1541–1547.

WELLS, PHILIP V.
    1965. Scarp Woodlands, Transported Grassland. Soils, and Concept of Grassland Climate in the Great Plains Region. *Science*, 148:246–249.

ZUCKERMAN, S.
    1957. The Human Breeding Season. *The New Scientist*, 1 (April) :12–14.

Book list

General Works on Primates

A HANDBOOK OF LIVING PRIMATES, by J. R. Napier and P. H. Napier. London and New York: Academic Press, 1967.

A HISTORY OF THE PRIMATES, by W. E. Le Gros Clark. 9th edition. London: British Museum, 1965.

THE LIFE OF PRIMATES, by Adolph H. Schultz. London: Weidenfeld and Nicolson, 1969.

MAN'S POOR RELATIONS, by Earnest Hooton. New York: Doubleday and Doran, 1942.

THE MONKEY KINGDOM, by Ivan T. Sanderson. Philadelphia and New York: Chilton Books, 1957.

THE OLD WORLD MONKEYS, edited by J. R. and P. H. Napier. New York: Academic Press. In press.

THE PRIMATES, by Irven DeVore. London: Seymour Press, 1969.

THE PRIMATES, edited by J. R. Napier and N. A. Barnicot. Symposia of the Zoological Society of London, Number 10, 1963.

PRIMATES, COMPARATIVE ANATOMY AND TAXONOMY (6 Volumes. 1953–1966), by W. C. Osman Hill. Edinburgh: Edinburgh University Press.

UP FROM THE APE, by Earnest Hooton. New York: Macmillan Company, 1947.

Books on Particular Primates

THE APES, by Vernon Reynolds. London: Cassell, 1968.

LEMUR BEHAVIOR, by Alison Jolly. University of Chicago Press, 1967.

MAN, by R. J. Harrison and W. Montagna. New York: Appleton-Century-Crofts, 1969.

MEN AND APES, by Desmond and Ramona Morris. London: Hutchinson, 1966.

THE MOUNTAIN GORILLA: ECOLOGY AND BEHAVIOR, by G. B. Schaller. University of Chicago Press, 1963.

MY FRIENDS THE WILD CHIMPANZEES, by Jane van Lawick-Goodall. Washington, D.C.: National Geographic Society, 1968.

THE SQUIRREL MONKEY, edited by Cooper and Rosenblum. New York and London: Academic Press, 1968.

THE YEAR OF THE GORILLA, by G. B. Schaller. University of Chicago Press, 1964. [A popular version of Schaller, 1963.]

Primate Behavior

NATURALISTIC BEHAVIOR OF NON-HUMAN PRIMATES, by C. R. Carpenter. University Park: Pennsylvania State University Press, 1964.

PRIMATE BEHAVIOR, edited by I. DeVore. New York: Holt, Rinehart & Winston, 1965.

PRIMATE ETHOLOGY, edited by Desmond Morris. London: Weidenfeld and Nicolson, 1967.

PRIMATES, edited by Phyllis C. Jay. New York: Holt, Rinehart & Winston, 1968.

SOCIAL BEHAVIOR FROM FISH TO MAN, by William Etkin. Phoenix Books. University of Chicago Press, 1967.

SOCIAL COMMUNICATION AMONG PRIMATES, edited by Stuart A. Altmann. University of Chicago Press, 1967.

Evolution of Man

THE ANTECEDENTS OF MAN, by W. E. Le Gros Clark. Edinburgh University Press, 1959.

ARBOREAL MAN, by F. Wood Jones. London: Edward Arnold, 1918.

CLASSIFICATION AND HUMAN EVOLUTION, edited by S. L. Washburn. London: Methuen, 1964.

EARLY MAN, by F. Clark Howell. London: Seymour Press, 1967.

THE EMERGENCE OF MAN, by John E. Pfeiffer. New York: Harper & Row Publishers, Inc., 1969.

EVOLUTION OF MAN, by David Pilbeam. London: Thames & Hudson, 1970.

EVOLUTIONARY AND GENETIC BIOLOGY. Vols. I and II, edited by John Buettner-Janusch. New York and London: Academic Press, 1963.

FOSSIL EVIDENCE FOR HUMAN EVOLUTION, by W. E. Le Gros Clark. 2nd Edition. University of Chicago Press, 1955.

A GUIDE TO FOSSIL MAN, by M. H. Day. London: Cassell, 1965.

HUMAN VARIATIONS AND ORIGINS. Readings from Scientific American. San Francisco, California: W. H. Freeman and Company, 1967.

MAN-APES OR APE-MEN, by W. E. Le Gros Clark. London: Holt, Rinehart & Winston, 1967.

MANKIND EVOLVING, by T. Dobzhansky. London: Yale University Press, 1962.

MANKIND IN THE MAKING, by William Howells. London: Pelican Books, 1967.

MAN'S EVOLUTION, by C. L. Brace and M. F. Ashley Montagu. London: Collier-Macmillan, 1965.

THE ORIGIN OF RACES, by Carleton S. Coon. London: Jonathan Cape, 1963.

ORIGINS OF MAN, by John Buettner-Janusch. Chichester: John Wiley and Sons, 1966.

PERSPECTIVES ON HUMAN EVOLUTION, edited by S. L. Washburn and Phyllis C. Jay. New York: Holt, Rinehart & Winston, 1968.

Evolutionary Theory

ANIMAL SPECIES AND THEIR EVOLUTION, by A. J. Cain. London: Hutchinson, 1966.

ANIMAL SPECIES AND THEIR EVOLUTION, by Ernst Mayr. London: Oxford University Press, 1963.

BIOLOGY AND MAN, by G. G. Simpson. New York: Harcourt, Brace and World, 1969.

CHARLES DARWIN: EVOLUTION AND NATURAL SELECTION, by Gavin de Beer. London: Nelson, 1963.

EVOLUTION, by Ruth Moore. Life Nature Library. London: Seymour Press, 1969.

THE LIVING STREAM, by Sir Alister Hardy. London: Collins, 1967.

ON THE ORIGIN OF THE SPECIES BY MEANS OF NATURAL SELECTION, by Charles Darwin. London: John Murray, 1859.

PRINCIPLES OF ANIMAL TAXONOMY, by G. G. Simpson. New York: Columbia University Press, 1961.

Human Behavior

AFRICAN GENESIS, by Robert Ardrey. London: Collins 1965; Fontana 1967.

THE HUMAN ZOO, by Desmond Morris. London: Jonathan Cape, 1969.

MEN IN GROUPS, by Lionel Tiger. London: Thomas Nelson, 1969.

THE NAKED APE, by Desmond Morris. London: Jonathan Cape, 1967.

ON AGGRESSION, by Konrad Lorenz. London: Methuen and Co., Ltd., 1966.

THE TERRITORIAL IMPERATIVE: A Personal Inquiry into the Animal Origins of Property and Nations, by Robert Ardrey. London: Collins, 1967.

VIOLENCE IN MONKEYS AND MAN, by Claire and W. M. S. Russell. London: Macmillan Company, 1968.

Index of authors

Index of subjects

Abominable Snowman, 70, 201
Adapidæ, 105
Adapis, 105
Adaptability of primates, 8, 14, 23, 30. ' ˜7
Adaptation, 24, 30
Adaptive radiation, 29
Aegyptopithecus, 107, 131, 190
Aeolopithecus, 107, 190
Aggressive behavior, 159, 220
Allometry, 19, 35, 123
Allopatry, definition of, 21
Allouatinæ, 51
Alouatta, 50, 51, 69
American Sign Language (A.S.L.) , 153
Anathana, 41
Angwantibo, *See Arctocebus*
Antigenic blood factors, 24
Anthropoidea, 48-64
Anthropology, 3
Aotinæ, 52
Aotus, 50, 52, 54, 109
Apes. *See also* Pongidæ; *Gorilla; Pan; Hylobates; Pongo.*
　evolution of, 60-64, 111-112, 131-137
　natural history of, 60-64, 119-125
Apidium, 107
Arboreal man, 70, 173
Arctocebus, 42, 43
Atavisms, 79-80
Ateles, 13, 50-51, 100, 135
Atelinæ, 50-51
"Aunt" behavior, 29
Australopithecus, 139, 141, 147, 193-198
　affinities, 194-197, 209-210
　age, 208
　characters, 194-197
　comparison with apes, 194-197
　comparison with *H. habilis*, 204, 206-207

discovery, 193
gait, 197
geological, 208
teeth, 196
tool-using, 204
Australopithecines. *See Australopithecus*
Avahi, 46-47
Aye-aye. *See Daubentonia*

Baboon. *See Papio*
Barbary ape. *See Macaca*
Bearded Saki. *See Chiropotes*
Behavior, evolution of, 27
Bigfoot, 70, 201
Bilophodonty, 61
Bipedalism, definition of, 98, 162-176
Birth season, 155
Birth weights, 129-130
Brachiation
　in chimpanzees, 121-122
　definition of, 97-98
Brachyteles, 66, 94
Brain, 48, 91, 126-127, 139, 151-152, 194, 205, 208
Bushbaby. *See Galago*

Cacajao, 50, 52
Callicebus, 52, 54, 109
Callimico, 50
Callithrix, 48, 50, 54
Capuchin monkey. *See Cebus*
Cascadian Revolution, 87
Catarrhine monkeys. *See Cercopithecoidea*
Cebidæ, 50-54
Cebinæ, 52
Ceboidea, 9, 10, 49-54, 90
Cebuella, 50
Cebus, 50, 52

235

GEORGE ALLEN & UNWIN LTD

Head office:
40 Museum Street, London, W.C.1
Telephone: 01-405 8577

Sales, Distribution and Accounts Departments
Park Lane, Hemel Hempstead, Herts.
Telephone: 0442 3244

Athens: 7 Stadiou Street, Athens 125
Barbados: Rockley New Road, St. Lawerence 4
Bombay: 103/5 Fort Street, Bombay 1
Calcutta: 285J Bepin Behari Ganguli Street, Calcutta 12
Dacca: Alico Building, 18 Motijheel, Dacca 2
Hornsby, N.S.W.: Cnr. Bridge Road and Jersey Street, 2077
Ibadan: P.O. Box 62
Johannesburg: P.O. Box 23134, Joubert Park
Karachi: Karachi Chambers, McLeod Road, Karachi 2
Lahore: 22 Falettis' Hotel, Egerton Road
Madras: 2/18 Mount Road, Madras 2
Manila: P.O. Box 157, Quezon City, D-502
Mexico: Serapio Rendon 125, Mexico 4, D.F.
Nairobi: P.O. Box 30583
New Delhi: 4/21–22B Asaf Ali Road, New Delhi 1
Ontario, 2330 Midland Avenue, Agincourt
Singapore: 248C–6 Orchard Road, Singapore 9
Tokyo: C.P.O. Box 1728, Tokyo 100–91
Wellington: P.O. Box 1467, Wellington